Springer Texts in Social Sciences

This textbook series delivers high-quality instructional content for graduates and advanced graduates in the social sciences. It comprises self-contained edited or authored books with comprehensive international coverage that are suitable for class as well as for individual self-study and professional development. The series covers core concepts, key methodological approaches, and important issues within key social science disciplines and also across these disciplines. All texts are authored by established experts in their fields and offer a solid methodological background, accompanied by pedagogical materials to serve students, such as practical examples, exercises, case studies etc. Textbooks published in this series are aimed at graduate and advanced graduate students, but are also valuable to early career and established researchers as important resources for their education, knowledge and teaching.

The books in this series may come under, but are not limited to, these fields:

- Sociology
- Anthropology
- Population studies
- Migration studies
- Quality of life and wellbeing research

Richard K. Thomas

# Demography: An Introduction to Population Studies

Springer

Richard K. Thomas
Memphis, TN, USA

ISSN 2730-6135 ISSN 2730-6143 (electronic)
Springer Texts in Social Sciences
ISBN 978-3-031-56622-6 ISBN 978-3-031-56623-3 (eBook)
https://doi.org/10.1007/978-3-031-56623-3



This Springer imprint is published by the registered company Springer Nature Switzerland AG
The registered company address is: Gewerbestrasse 11, 6330 Cham, Switzerland

# Preface

The publication of an introductory textbook on demography no doubt raises the question: Why another introductory demography book? That is a good question, and there are several reasons why a new book on demography makes sense at this time. Actually, there are relatively few basic introductions to demography on the market and, from this author's perspective, there is no existing work that captures the sense of contemporary population studies. Some of the extant works appear to have been written for other demographers and fail to provide the basic understanding of the field that should be the take-away for a beginning student. Virtually none exhibits the "look and feel" of a modern social science textbook, being heavy on text and light on graphics, illustrations and case studies. Further, these texts generally lack the hands-on activities that showcase demography as an applied science. Unless one actually "does" demography, it could be argued, it is impossible to appreciate the significance of the discipline.

The idea of a new textbook was also driven by the author's 50 years of experience teaching demography in the classroom and practicing demography in the community. In many ways, this is a book of "lessons learned" over a long career in the field. It reflects not only the personal academic and applied experiences of the author but also pays homage to those who have pioneered applied demography and made significant contributions to the diverse areas to which demography can be applied.

Another, and in many ways the most important reason, is that the world in which demographers operate and in which we as society members reside has changed significantly over the past twenty years and dramatically over the past three years. Looking at the long run it could argued that many of the assumptions that we as demographers have made are no longer valid. Certainly, the beliefs held by American citizens about our society are extremely dated and reflect an era that is long gone (if it ever existed!). In the short run, it should be noted that this work was completed while the COVID-19 pandemic was still underway. While we have to add qualifiers to much of the data presented in this book, it is clear that any textbook written before 2020 will inevitably be dated. Nothing is the same as it was, with the pandemic having a major impact on the American economy, education system, and healthcare system, among others. Indeed, current population statistics are a result of the impact of the pandemic, with its implications for fertility, morbidity, mortality, and migration. In fact, there is virtually no topic covered in demography that has not been affected by the pandemic. At the very least, the statistics in any recently published demography textbook would have to be updated.

One final rationale for adding an introductory demographic textbook to the collection of works currently available is the practical matter that the publisher, Springer Nature, the largest publisher of demographic works in the world, does not have an introductory demographic book in its extensive holdings. This fact and the justifications noted above suggest that this textbook is in the right place at the right time. Ideally, it will inspire those early in their college journeys to consider a career in demography. There is no better place to learn about the world and your place in it.

Because this work was primarily written during the COVID-19 pandemic and completed during the last stages of this worldwide health crisis the data presented run the risk of not being representative of a "normal" time period. For this reason, it was thought that figures for 2021 and 2022 would not be reflective of past trends resulting in their limited use. In any case, while the long-term impact of the pandemic on demographic trends has yet to be determined, some aspects of our societal mosaic will be significantly changed as a result of the pandemic.

Memphis, USA Richard K. Thomas

# Contents

CHAPTER 1

# An Introduction to Demography

**Abstract** This chapter defines and describes the field of demography (or population studies). It explains the demographic perspective and what it means to view the world through "demographic eyes." It describes what is unique about demography and how it is differentiated (and is akin to) the "hard" sciences and other social sciences. The idea of "demography as destiny" is explored, and the mutually influential aspects of demography and social trends are examined. Demography as an *applied science* is emphasized.

Demography as a distinct field of study (also referred to as "population studies") has a relatively short history. The term was first used in 1855 by Guillard (2010) when he published *Elements de Statistique Humaine ou Demographie Comparee.* Guillard coined the term based on a Greek word meaning "describing people"—"demo" for people and "graphy" to write about a particular topic. To Guillard the focus of demography was the mathematical knowledge of populations, their general movements, and their physical, civil, intellectual and moral state. His interest in population size and distribution, demographic processes and the structure of a population foresaw modern demography.

***The population characteristics of interest to demographers are those that are relevant within a social and cultural context.***

Although demographers are interested in the characteristics of populations, they are not interested in every characteristic. There is a particular set of attributes that are the focus of demographic analysis. The characteristics of interest to demographers are those that are relevant within a social and cultural context. Thus, demographers study biosocial traits such as age, sex and race



R. K. Thomas, *Demography: An Introduction to Population Studies*, Springer Texts in Social Sciences, https://doi.org/10.1007/978-3-031-56623-3_1

and sociocultural traits such as marital status, education, income, occupation and even religion.

Because demography is a dynamic science, an understanding of demographic trends is important. Changes in birth rates, death rates and patterns of migration are of major importance to demographers. They represent the "moving targets" of any population and not only provide a snapshot of where we are now but of where we as a people are going. These trends in turn offer an indication of anticipated changes in the attributes of any population under study.

Demographers do not focus on the attributes of individuals but on the characteristics of groups of people. While every individual might be considered to possess a "demographic profile", the interest of demographers is on the attributes of aggregates—a community, a state or a nation. There are situations in which a subset of a population may be the object of study, such as cohorts of child-bearing age women, senior citizens, or African-Americans, but it is still the aggregate characteristics of the group that are of significance. There will be significant variation within any group in terms of its attributes, but it is the "average" characteristics that are of interest to the demographer.

***Demographers do not focus on the attributes of individuals but on the characteristics of groups of people.***

## Demography as an Applied Science

Demography is by definition an *applied* science. While this course devotes some time to demographic theory, most of the material is devoted to the study of the concepts, methods and data used for applications of demography to real-world problems. A frequent query from demography students is: This is interesting but what can I do with it? Like all disciplines, there is a basic science dimension to demography, what we might call science for science's sake. Ultimately, however, most demographers use their knowledge to understand real world problems and, not just to understand them, but to help develop solutions for them. Demographers analyzing the changing age structure of the US population, for example, have contributed important knowledge for addressing the challenges of funding the federal Social Security and Medicare programs. Demographers examining fertility trends have contributed to solutions for addressing such phenomena as America's relatively high infant mortality rate.

***Demography is by definition an applied science, and demographers use its techniques to address real-world problems.***

In addition to proposing practical solutions to real-world problems, demographic data make an important contribution to policy setting. As in the case of Social Security and Medicare cited above, an understanding of demographic trends provides policy-makers with the background they need for establishing effective policies. There is essentially no sector of US society that cannot benefit at the policy level from demographic input. Thus, education, economic development, transportation, disaster preparedness, and criminal justice, to name a few, are areas where demographic data can make a significant contribution to policy setting and implementation.

## Why Study Demography?

The study of demography is important for a number of reasons. For starters, nearly everything is connected to demography (Weeks, 2008). Demography describes "our world"—and description is the starting point for understanding it and, ultimately, taking action to improve it. "Our world" could be any collection of people with which we are involved—a social group, classroom, neighborhood, city, or the whole world for that matter.

The relevance of demography for an understanding of the world is reflected in the major issues making headlines in recent years:

- Record inflation affecting the US economy
- The unexpected increase in the death rate for Americans
- The effects of climate change on various parts of the world
- The continuing issue of illegal immigration into the US
- The impact of the COVID-19 pandemic on population processes

These issues all have national and/or international implications—and all are directly or indirectly related to demography. In fact, there is virtually no social, economic or political issue that does not have its roots in the demographics of the population.

While the events above have captured the headlines, there are a number of other trends occurring within the US population that reflect long-term changes in the social structure:

- The aging of the US population (accompanied by growing "feminization")
- The increasing racial/ethnic diversity of the US population
- The changing family structure (now including same-sex marriages)
- The declining economic status of some segments of American society
- The changing disease patterns within the US population

These types of demographic trends have significant implications for US society present and future, and there is no social institution that is not impacted by these developments. The aging of the US population by itself has ushered in an unprecedented era for a society that has always emphasized its youthfulness. The dramatic increase in the Hispanic population in the US has significant implications for our economy, education, healthcare and politics.

While these major trends related to demographics affect society as a whole, there is a personal dimension to this as well. As Weeks (2008) points out, the demographic foundation of our lives is deep and broad. Although demographers are interested in the characteristics and behavior of groups of people, the demographic attributes of our society affect nearly every aspect of our personal lives in one way or another. The types of personal decisions that we as individuals make have a cumulative effect on population trends. Some of the decisions that affect our daily lives are:

- The decision to get married (or not) and when
- The decision to have children (or not) and when
- The neighborhood we choose to make our home in
- The type of job we choose or are looking for
- The educational level we aspire to
- The choice of political party to support
- The health-related behaviors we carry out

Thus, large numbers of individuals deciding not to get married means that, today, we have the lowest marriage rate in our history. Or, large numbers of people deciding not to have children has resulted in a reproduction rate so low that we are not able to maintain our population size. The cumulative actions of individuals ultimately serve to create the trends that change the nature of society.

***Although demographers are interested in the characteristics of groups of people, the demographic circumstances in our society affect nearly every aspect of our personal lives.***

Although we often act of our own volition to make decisions that not only affect our lives but the demographic trends that affect our society, we must remember that the impact demographic factors have on society represents a "two-way street". On the one hand, our population attributes influence the demographic processes characteristic of the society. On the other hand, the demographic processes operating in society affect our demographic attributes. Thus, while the age structure significantly determines the birth rate, the birth significantly determines the age structure.

Everyone within a population was essentially born into a game that was already under way. The rules were written long before we were born, so the type of education we aspire to (and our ability to achieve it), the type of job that we are able to obtain, the type of person we are likely to marry, and even the political party we are likely to support, are all reflections of our position within the social structure. Sadly, someone born into a low-income, minority household in the South is probably "doomed by demographics". They did not choose this situation, but the circumstances of their birth will probably follow them throughout their lives. A recent study, in fact, reported that the best predictor of a healthy life is the ZIP Code where you live (Roeder, 2014). While members of society are able to control their demographic futures to a certain extent, we are all constrained by the demographic parameters surrounding our position in society. Thus, we are both the "creators" of and the "victims" of our demographic circumstances.

All of this means there is a strong correlation between demographic characteristics and certain attributes and behaviors. In fact, some demographers make a living projecting the future characteristics of populations based on what is known about that population's current demographic makeup. It would not be unusual, for example, for analysts to use demographic methods to predict how long members of a certain demographic group are going to live and, in fact, determine prospectively what diseases they are likely to die from, or for demographers to predict election results based on the demographic characteristics of likely voters. The bottom line is: Knowing the demographic characteristics of a population opens a door to an understanding of a wide variety of attributes of that population, some of them seemingly unrelated to demographics.

***Knowing the demographic characteristics of a population opens a door to an understanding of a wide variety of attributes of that population.***

Ultimately, through our individual behaviors we make our demographic bed and then have to lie in it. The society we have created through our behaviors, for example, has the following implications for our lives:

- Our population is no longer growing
- We are getting steadily older as a society
- We have fewer "traditional" families
- We have a predominantly female population
- We are becoming sicker as a population

As we consider these trends, one of the things that becomes obvious is that societies are constantly undergoing change—some slower, some faster—but change is always occurring. In the US, we have made a fetish of change and virtually incorporated demographic change into the DNA of American society. No other society has experienced as much change as we have over such a prolonged period of time. We change jobs, houses, spouses and lifestyles with startling frequency. This preoccupation with change has significant implications for demographic trends.

***Societies are constantly undergoing change, whether as a result of some traumatic event or simply through a slowly changing demographic structure.***

The changes that have been occurring in US society are so significant that we are constantly setting new records. Today, for example, the American population contains:

- The largest elderly population we have ever had (both numbers and percent)
- The greatest "excess" of women we have ever had
- The lowest marriage rate we have ever had
- The largest group of single adults we have ever had (both numbers and percent)
- The largest number of single-parent households we have ever had
- The most women working outside the home we have ever had
- The highest level of immigration (both legal and illegal) we have ever had

The magnitude of change occurring in US society is a reflection of the contribution of human behavior to the demographic trends that are underway. At the same time, these developments contribute to significant changes in the characteristics of the American population. As a result, our American "portrait" in the early twenty-first century looks much different than it did only a couple of decades ago.

## Who Uses Demography?

People in every sector of society employ demographic data, often without realizing that is what they are doing. Increasingly we hear people speak of the "demographics" of a consumer group or the fact that a certain "demographic" always votes for a certain political party. Even though public expressions about demographics are becoming more common, the widespread use of demographic data and methods is not widely appreciated.

Members of the business community, particularly those involved in marketing, pioneered the use of demographic data in the private sector. Virtually no business decision is made in the corporate boardroom today without considering the relevant demographics. Whether the decision involves identifying a target audience for a new product, determining the location for a new store, or designing a sales territory, the demographics of the population under question are a critical piece of the puzzle. This process does not just relate to major corporate decisions but affects us as individuals. The fact that we receive certain catalogues, certain types of junk mail, or certain telephone solicitations reflects the information that marketers have about our demographic attributes.

While the business community was the first to recognize the importance of demographics, its representatives are far from the only users of such data. In every aspect of American life demographics have become increasingly important. The allocation of government services depends on an in-depth understanding of the characteristics of the population to be served. Indeed, the original intent of the census conducted by the federal government every ten years was for the apportionment of Congressional districts. Those who aspire to political office begin their campaigns with an assessment of the demographics of their prospective constituents.

***While the business community was the first to recognize the importance of demographics, people in every sector of society employ demographic data.***

Those involved in urban planning and community development start with the demographics of the geographic area under consideration. The education system depends on an understanding of the number, location and characteristics of school-aged children, and the services provided by the healthcare system are a direct reflection of the characteristics of the patient population. The military must plan to accommodate the characteristics of potential inductees, and, clearly, today's American armed forces reflect the changing demographic character of our society with record numbers of women and members of various racial and ethnic groups.

In most of these examples, demographics are put to good use—good for the users and good for society. But there are cases where less well-intended individuals and organizations may use demographics for less than noble purposes. Much has been made of racial and ethnic profiling used by law enforcement agencies and championed by those who oppose immigration into this country. Certain groups may be discriminated against due to their demographic characteristics—their race or ethnicity, income level, religion or language. Historically, individuals have been excluded from housing developments, social clubs and occupations due to their sex, race or cultural background. As with any aspect of the human condition, demographics can be used for good or ill.

## Demographic Perspectives and Methods

An understanding of the "demographic perspective", the concepts that define the discipline, the methods used for demographic analysis, and the data utilized in these analyses is essential. The demographic perspective involves a way of seeing the world within a demographic context or through a "demographic lens". This unique manner of viewing the world provides a framework for interpreting social phenomena and a means of linking social groups to their environments. This perspective leads to a search for interrelationships between demographic variables and other variables associated with a population.

Demographers focus on a number of different processes that are inherent to a population's demographic trajectory. The key demographic processes—fertility, mortality and migration—represent the three dimensions of population change. Demographers have recognized other social developments that effect these key processes such as the "demographic transition" and the "epidemiological transition". The latter considers the emergence of morbidity as an important "process" that affects a population.

***The three key demographic processes are fertility, mortality and migration.***

Demographers have also posited a number of theories that provide a framework for analyzing developments in society. These include theories of population change, growth and decline, theories related to the relationship between human populations and their environments, and theories that address the interaction of various demographic variables within a population.

Demographers also utilize a variety of methods to analyze populations and determine their characteristics. They use *cohort analysis* to track segments of a population (e.g., baby boomers) over time and to predict future population trends. They use *spatial analysis* to examine the relationship between populations and the physical environment. *Epidemiological analysis* is utilized to examine and explain the distribution of disease within a population. All of these analytical techniques are supported by well-developed methods for estimating and projecting the size and characteristics of a population.

One conclusion generated by looking at the population through a "demographic lens" is that virtually no demographic or social attribute is randomly distributed within society. All social phenomena (including demographics) display patterns. When demographers look at a population, they do not see an amorphous mass of individuals, but they see clusters of characteristics and patterns reflecting the distribution of various demographic traits. Look at any neighborhood, for example. Are the residents randomly distributed within that community from the general population? Not likely—they are going to exhibit certain traits of age, household structure, educational and income levels, lifestyles and so forth. We know now that even illness is not distributed randomly but exhibits very clear patterns within a population.

***Demographers utilize a variety of methods to analyze populations and track their characteristics over time.***

## Major Topics in Demography

### *Population Size and Distribution*

The first facts that a demographer looks for related to a population are the number and distribution of the people resident therein. The size of the population is a key demographic variable and provides the framework for additional analysis. Equally important, however, is the distribution of that population across the geography in question. In the US, for example, how are our 335 million residents distributed over various states and among various cities? To what extend is our population widely dispersed (e.g., in New Mexico) or tightly concentrated (e.g., in New York City)? These questions can be raised for any geographic or political unit and the nature of the distribution of the population has significant implications for its demographic profile. At the same time, the population's demographic makeup will contribute to the manner in which the population is distributed. Thus, in the US we find American Indians (a racial group) primarily isolated on "reservations" in the western states, African Americans often segregated within our urban areas, and senior citizens concentrated in popular retirement locales (such as Florida and Arizona).

***The size of the population is a key demographic variable and provides the framework for additional analysis.***

Demographers are not only interested in the current size and distribution of the population but in future attributes as well. For most practical purposes, it is important to develop an appreciation for the size and distribution of the population ten, twenty or even more years into the future. To this end, demographers have developed a variety of techniques for generating projections and forecasts of the size and distribution of future populations.

### *Demographic Processes*

Historically, demographers have emphasized the study of the major demographic processes—fertility, mortality and migration—and a significant portion of the book is devoted to these topics. Fertility, which refers to the process of reproduction and childbirth, is the primary way in which new members are added to a group. Thus, the number of births and the characteristics of those births are of particular interest to demographers. Mortality refers to the study of death and the characteristics surrounding death within a particular society. This is the primary means through which people are subtracted from a group, and the mortality characteristics of a particular society tell us a lot about that society. Migration—the process through which individuals, families and groups move from one place to another—is the third process to be considered. Migration can involve internal movement (e.g., within the US) or international movement (i.e., from country to country). Obviously, migration can both add and subtract people from a population, and, in today's world, this process often has more implications for society than births or deaths.

Another process of growing interest to demographers is the morbidity patterns associated with a population. Morbidity analysis focuses on the health problems that characterize a population and the distribution of these conditions across society. A population's morbidity characteristics are subject to change and have important linkages to patterns fertility and mortality.

These three key processes taken together are important in that they are the components of population change. The size of a population is changed through a combination of births and in-migration adding members while deaths and out-migration subtract members. As will be demonstrated later, the population in Time 2 is a result of the population in Time 1 plus births minus deaths plus/minus migration. Understandably, demographers place a lot of attention on these processes due to their impact on the size and characteristics of any particular population.

### *Population Composition*

Another major topic to be covered in this text is the demographic characteristics associated with human populations. These are the attributes that give a population its "character" and distinguish one population from another. For our purposes we will divide demographic characteristics into biosocial traits such as sex, age and race and sociocultural traits such as marital status, income, occupation, and education. These attributes give our society its structure and determine much of our behavior. A population's demographic composition influences the key demographic processes, while, at the same time, the operation of these demographic processes influences the population's composition. When demographers examine population trends, both the change in the number of people and changes in the population's composition are important. An understanding of biosocial and sociocultural attributes makes it possible to profile a population in terms of its key demographic characteristics.

## Organization of the Book

The material in this text is organized in such a manner as to systematically introduce the materials and methods of demography to the student. Chapter 2 provides an overview of the demographic approach to the study of human populations. It describes the perspectives, concepts, and methodologies commonly used by demographers in the pursuit of their craft and further distinguishes the demographic method from other modes of study of human populations. Chapter 3 examines the basic demographic concepts of population size, distribution and concentration. This chapter also introduces the geographical units used by demographers as a context for analysis.

Chapters 4 and 5 focus on population composition and introduce biosocial attributes and sociocultural attributes, respectively. Chapter 4 considers the population attributes of age, sex and race and ethnicity, often considered core demographic traits. Chapter 5 examines the more diverse set of attributes referred to as sociocultural traits, including marital status, educational level, income level, and occupation among others. Chapter 6 is devoted to a discussion of the data that are so basic to demographic analysis. The types of data used by demographers and how to access them are thoroughly considered.

The next three chapters, Chaps. 7–9, consider the three basic demographic processes: fertility, mortality and migration. Chapter 7 examines fertility and the primary way in which new members are added to a group. The number of births and the characteristics of those births are of particular interest to demographers. Mortality, addressed in Chap. 8, refers to the study of death and the characteristics surrounding reported deaths, with death being the primary means through which people are subtracted from a group. In Chap. 9, we take a close look at migration—the process through which individuals, families and groups move from one place to another. Obviously, migration can both add and subtract people from a society and, often has more implications for changes in population size and characteristics than either births or deaths.

Chapter 10 addresses the concept of population change and the factors that contribute to population growth and decline. Particular attention is paid to the generation of estimates and projections on the size and composition of a population. These types of computations (synthetic data) are important for generating estimates and projections when actual data are not available.

Chapters 11 and 12 essentially summarize the material presented in the previous ten chapters by describing the state of the populations of the United States and the world with emphasis on the ongoing trends characterizing the respective populations. Chapter 11 looks at the current demographic profile of the United States and describes the trends that are changing the demographic landscape. Chapter 12 follows the same procedure for the world, profiling the current status of the world's population and describing the trends that are affecting it. In both cases, predictions of future population patterns will be offered.

Chapter 13 presents an overview of applied demography. As previously noted, demography is by definition an applied science. This chapter describes the way in which demographic data, concepts and methods are applied to real-world problems, examining applications in business, healthcare, education and community development among others. Examples are provided of concrete applications of demographic methods in various spheres of society.

Hands-on activities are provided throughout the textbook to provide students exposure to the application of demographic data and methods to concrete problems. A glossary is provided at the end of the textbook to allow beginning students of demography to understand the basic concepts and terms used by demographers.

## Key Points

- "Demography" comes from Greek words that mean "describing people".
- Demographers do not focus on the characteristics of individuals but on those of groups of people.
- Demography is an applied science, and most demographers use demographic methods to address real-world problems.
- An understanding of demography is important for understanding developments that occur within the world's populations.
- There is virtually no social, economic or political issue that does not have its roots in demography.

- Although demographers are interested in the characteristics of groups of people, the demographics of our society affect nearly every aspect of our personal lives.
- The impact of demographics on society represents a "two-way street"—with demographic attributes affecting demographic processes and demographic processes affecting demographic attributes.
- Knowing the demographic characteristics of a population opens the door to understanding many other attributes of that population.
- Societies are always changing whether through a slow, incremental process or as a result of some catastrophic change.
- While the business community was the first to recognize the importance of demographics, people in every sector of society employ demographic data today.
- The three key demographic processes are fertility, mortality and morbidity.
- The first demographic "fact" that a demographer considers is the size of a population.

## References

Guillard, A. (2010). *Elements de Statistique Humaine ou Demographie Comparee*. Nabu Press.

Roeder, A. (2014). *Zip code better predictor of health than genetic code*. http://www.hsph.harvard.edu/news/features/zip-code-better-predictor-of-health-than-genetic-code/

Weeks, J. R. (2008). *Population: an introduction to concepts and issues*. Thomson Wadsworth.

## Additional Resources

Lundquist, J. H., Anderton, D. L., & Yaukey, D. (2014). *Demography: The study of human populations* (4th ed.). Waveland Press.

Swanson, D. A., & Siegel, J. S. (2004). *The methods and materials of demography* (2nd ed.). Emerald Group Publishing.

Singelmann, J., & Poston, D. L. (Eds.). (2020). *Developments in demography in the 21st century*. Springer.

Yaukey, D., Anderton, D. L., & Lundquist, J. H. (2007). *Demography: The study of human populations*. Waveland Press.

CHAPTER 2

# The Methods and Materials of Demography

**Abstract** This chapter reviews the theories, methods and data utilized by demographers in their demographic pursuits. Although demography as an applied science does not emphasize theory, there are some key theories that have particular relevance for demographers. More importantly, this chapter reviews the various concepts that have been developed by demographers and the methods that are utilized in pursuit of solutions to concrete problems. An introduction to the data used by demographers is offered. This chapter emphasizes the fact that virtually no attribute is randomly distributed within society and highlights the influence of demographic factors on most aspects of society.

## Introduction

Demography is distinguished from other fields of study by virtue of its unique set of attributes. While sharing intellectual space with a number of other academic disciplines, demography has developed perspectives, concepts and theories that set it apart. Many of these attributes were originally catalogued in the classic demographic work, *The Methods and Materials of Demography* (Swanson & Siegel, 2004). The sections that follow describe the various dimensions of demography that are unique to the field while recognizing contributions made by other disciplines.



R. K. Thomas, *Demography: An Introduction to Population Studies*, Springer Texts in Social Sciences, https://doi.org/10.1007/978-3-031-56623-3_2

## The Demographic Perspective

The demographic perspective provides a unique way of seeing the world. When a demographer views the world through a "demographic lens" he sees things that others do not. The "world" of course could mean literally the entire world or any population of interest. When demographers observe a classroom of students, the residents of a small town, or the citizens of the United States, for example, they do not see an amorphous mass of individuals, but they see a well-ordered population with more or less sharply delineated features.

Societies are complex entities, and the United States is perhaps the most complex of all. Demography provides us a tool for making sense of this complexity. With our demographic lens we can segment the population we are observing into meaningful subgroups. Through this process we begin to gain understanding into the nature of that population. This understanding allows the observer to gain insights into relevant behaviors observed within that population.

Much of this text is devoted to a discussion of what demographers see when they look at a population from a demographic perspective. Some of the more obvious things that demographers observe are distinctions based on age, sex and race. All of us, of course, intuitively notice such basic characteristics for a group that we are observing. We note whether it is a young crowd or an old crowd, the ratio of men to women, and the extent to which the population is racially mixed. Demographers notice these same attributes either directly through observation or indirectly through compilation of statistics.

***Demography provides a tool for dividing a population into meaningful subgroups for further analysis.***

Demographers take these observations and attribute a wide range of other characteristics to the population under study. If we know the age composition of the population in question, for example, we know a lot of other things about that population. We can make assumptions about the population's family structure, its lifestyles and even its health status. In fact, health demographers can develop a profile of the health problems of a population with startling accuracy just by knowing the age structure. Demographers also know that a young population is typically going to have a majority of males and that an old population is going to have a majority of females. By knowing the age and sex distribution of the population, demographers can draw conclusions concerning marital status and household structure. By examining combinations of demographic characteristics, it becomes possible to conceptulize occupational status, educational attainment, income levels and so forth.

It is possible for demographers to make these assumptions concerning the population under study because the various demographic attributes are highly interrelated. People of a certain age, for example, are likely to have attributes that set them apart from people in other age cohorts. For a variety of reasons, people who are members of a certain racial or ethnic group are likely to have characteristics that set them apart from members of other racial and ethnic groups. This is not to stereotype members of various groups, of course, and there are always going to be individual exceptions. It does, however, underscore the fact that demographic attributes are interrelated and tend to be distributed in clusters throughout a population.

As noted previously, demographics is a two-way street, with the demographic characteristics of a population determining its behavior while, at the same time, the behavior of the population ultimately determining its demographic characteristics. These relationships are particularly clear when one examines basic demographic processes. Of particular significance is the impact of demographic characteristics on fertility or the reproductive behavior of the population.

***Demography provides observers with a means of understanding social phenomena.***

## Demography and Freewill

From a strict demographic perspective US society places an inordinate emphasis on individualism and often assumes that people act of their own volition. However, it could be argued that one's social circumstances (manifested in demographic attributes) is a greater predictor of fertility patterns, for example, than the vagaries of personal choice. While there are clearly childbearing-aged members within low-income minority groups in the United States who do not want to have children, the fact of the matter is that the likelihood of pregnancy for this segment of the population is much higher than that for virtually any other group.

While reproduction is a complicated process and a lot of factors come into play, it is clear that low-income populations have higher fertility rates than high-income populations, that certain racial and ethnic groups have higher fertility levels than other racial and ethnic groups, that immigrant groups have higher fertility rates than the native-born population, that the poorly educated have higher fertility rates than the well-educated, and so forth.

Demographers are reluctant to assert that "demographics is destiny," but it is clear that one's demographic traits exert a considerable influence on reproductive behavior. Because of these clear-cut patterns, demographers are able to predict future fertility patterns and, perhaps more important, predict the future characteristics of the population. Demographers know, for example, based on this information that the native-born white population is going to continue to shrink as a proportion of the population for the foreseeable future while minority populations continue to increase their proportion.

At the other end of the lifespan, mortality rates are influenced by demographics. Except in rare circumstances, individuals do not choose the time or cause of their deaths. If anything, it would be assumed that death is a somewhat unpredictable event and that mortality patterns are more a function of biological threats, environmental toxins and accidents among other causes that have little to do with either individual choice or demographic attributes. While the relationship between demographic attributes and mortality is not as clear cut as that for fertility patterns, the demographic characteristics of any population serve as reasonable predictors of both the level of mortality and the most common causes of death. In fact, the correlation between certain demographic attributes and the risk of death is so high that it is possible to calculate the odds of dying for any age group within US society.

Further, the cause of death for members of a particular age group can also be anticipated. As with fertility, membership in various demographic categories is associated with risk of death. We find, for example, that males are at much greater risk of death than females (and in every age cohort), that African Americans are at much greater risk of death than whites, that members of poverty populations are at much greater risk of death than those in affluent groups and so forth. Because of these patterns, demographers are able to predict future mortality levels and the leading causes of death years in advance.

***Knowledge of the demographic attributes of a population allows demographers to infer a lot about its other characteristics.***

A couple of other examples should suffice to underscore the need to understand demographics in order to interpret national and international events. Climate change has been increasingly in the news, and there is now conclusive evidence that human behavior is playing a major role in this phenomenon. A large part of the blame is placed on the economic activity of more industrialized nations. These activities are driven by demographically generated demand for energy and the goods that energy produces. In less developed parts of the world, a lot of the pressure on the environment is driven by a growing population which, based on numbers alone, requires ever increasing resources. Add to this the fact that the middle classes in many less-developed countries are growing at an unprecedented rate and there is a perfect storm for exploiting the environment and inducing climate change. If human behavior is driving climate change, demographic change is driving that behavior.

***The demographic perspective provides a means for linking social groups to their environments.***

The demographic perspective provides a means for linking social groups to their environments. Virtually all demographic data are linked to some geographic unit thereby connecting that population with the particular characteristics of that geography. Knowing that a population resides in the inner-city, the suburbs or a rural area has important implications for our understanding of that population and its attributes. Similarly, knowing that a population lives in an impoverished urban community rather than an affluent suburb, in single-family detached housing rather than high-density apartments, or in an isolated ethnic enclave rather than a diverse community also has implications for that population's characteristics. By segmenting a population on the basis of its geographic distribution, its type of community and its salient demographic features, demographers can establish a context for the future analysis of that population.

## Predicting Demographic Trends

One final aspect of the demographic perspective is the emphasis placed on the ability to predict demographic trends. Demographers have advantages over those in many other disciplines for a couple of reasons. First, as indicated above, the interrelationships between demographic variables are such that it is no great leap to be able to project the future characteristics of a population based on known current characteristics.

If one knows, for example, the current age and sex distribution for a population, it is possible to project the fertility and mortality patterns for that population ten or twenty years into the future. Of course, the analyst is making certain assumptions about fertility and mortality rates and, even more risky, about the behavior of individuals. Even so, such projections are often made with surprising accuracy. If we add other variables like marital status to the mix, the ability to predict future fertility levels is enhanced. These predictions could be further refined by profiling the population under study in terms of its racial makeup or socioeconomic status.

One other way in which demographers have an advantage is by virtue of the natural aging process exhibited by a population. In the absence of high levels of in-migration or out-migration and with some knowledge of reproduction rates, it is possible to project characteristics of a population into the future based on age distribution alone. For example, demographers can project the number of children entering kindergarten in five years because those children are already in the population (i.e., members of the population one year old or less). Similarly, demographers can project the number of youth entering college ten years from now because they know the number of eight-year-olds currently in the population. Or, demographers can project the number of new "senior citizens" that will be added to the population thirty years from now because they know how many 35-year-olds are currently in the population (as well as the number of those who are likely to die over the next 30 years).

***Demographers have an advantage because of their ability to project the characteristics of a population into the future based on age distribution alone.***

Although in each case the demographer has to make certain assumptions (e.g., college admission rates, mortality rates), the natural aging of the population allows the analyst to use standard demographic methodologies to project the future characteristics of the population.

## Basic Demographic Processes

Historically, demographers have emphasized the study of the major demographic processes—fertility, mortality and migration. *Fertility*, which refers to the process of reproduction and childbirth, is the primary way in which new members are added to a group. Thus, the number of births and the characteristics of those births are of particular interest to demographers. Fertility is an important topic for demographers because of its role in population change and the fact that differential fertility rates contribute to a changing population composition.

***The fertility patterns characterizing various groups within society will influence that population's future demographic traits.***

*Mortality* refers to the process of dying, and mortality analysis concerns the characteristics that surround deaths within a particular society. The mortality characteristics of a particular society tell us a lot about that society, not only about its health problems (morbidity patterns) but about society's social structure and culture. The changing cause-of-death patterns are of particular interest to demographers because of their implications for the health of the population.

*Migration*—the process through which individuals, families and groups move from one place to another—is the third process to be considered. Migration can involve internal movement (e.g., within the US) or international movement (i.e., from country to country). Obviously, migration can both add and subtract people from a society and, in today's world, often has more implications for society than births or deaths. Importantly, migration more than any other process influences the demographic characteristics of both the sending community and the receiving community. The facts that the population of Florida is relatively old and the San Antonio's population includes more Mexican-Americans than native-born whites reflect the impact of migration.

One emerging process that will be considered to a lesser extent relates to the morbidity of a population. *Morbidity* refers to the health problems that characterize a population and this topic is of increasing interest to students of demography. The study of mortality has become less important as death rates have declined thereby reducing the role of mortality in population change. The morbidity characteristics of a population have important linkages directly to mortality and indirectly to fertility and are both a reflection of and a contributor to changing demographic characteristics.

Fertility, mortality and migration taken together are important in that they are the components of *population change*. The size of a population is affected through a combination of births and in- migration adding new members and deaths and out-migration subtracting existing members. As will be demonstrated later, the population at the end of the period in question (Time 2) is a result of the population at the beginning of the period (Time 1) plus births minus deaths plus or minus migration. Understandably, demographers place a lot of attention on these processes due to their impact on the size and characteristics of any particular population.

***Fertility, mortality and migration taken together are important in that they are the components of* population *change.***

## Demographic Theories

Demographers (and others) have developed a variety of theories that focus on demographic processes and their interrelationship with demographic characteristics. Although theories are not as prominent within the demographic discipline as in most other academic fields due to the applied nature of demography, there are some theories that might be presented as *examples* of frameworks used to explain certain demographic phenomena.

### *Theories of Population Growth*

The variations displayed by populations as they grow and change has always fascinated demographers, and several theories put forth by demographers address the process of population growth.

Malthusian theory. The essay on the principle of population growth, an important work of Thomas Robert Malthus, a nineteenth century British economist, is a landmark in the history of population studies. He argued that the tendency of the population to grow faster in relation to its means of subsistence has led to human misery and placed several obstacles in the path of human progress Malthus observed that an increase in a nation's food production improved the well-being of the population, but the improvement was temporary because it led to population growth, which in turn restored the original per capita production level. In other words, humans had a propensity to utilize abundance for population growth rather than for maintaining a high standard of living. Populations had a tendency to grow until the lower class suffered hardship, want and greater susceptibility to war, famine and disease.

Malthusian theory stood in opposition to the popular view in 18th-century Europe that saw society as improving and in principle as perfectible. Malthus saw population growth as inevitable whenever conditions improved, thereby precluding real progress towards a utopian society.

Neo-Malthusian theory. Neo-Malthusians maintain that, although the gloomy predictions of Malthus may have been premature, they are basically correct. According to neo-Malthusians, the world's resources are adequate for a much larger population. Exploitation, not overpopulation, is the basic cause of world hunger. Neo-Malthusianists believe that population control through the use of contraception is essential for the survival of the earth's population. Their argument rests on the observation that resources are limited, and that growing populations could rapidly outstrip the provision of resources including land and food.

Optimum population theory. According to the supporters of this theory, populations grow up to certain desirable level after which further growth is harmful. The two important principles of this theory are:

1. When there is an increase in population the ratio between the total population and the working population must remain almost constant.
2. When at a point of time the population of a country increases, the natural resources, capital and technical know-how must change or the law of diminishing returns begins to operate.

## Demographic Methods

### *Demographic Analysis*

Demographic analysis is a technique used to develop an understanding of the age, sex, and racial composition of a population and how it has changed over time through the basic demographic processes of birth, death, and migration. Demographic analysis allows us to measure the dimensions and dynamics of populations. These methods have primarily been developed to study human populations and examine how populations can change across time through the processes of birth, death and immigration.

***Demographers use a variety of techniques for studying population characteristics and processes.***

### *Estimates and Projections*

Demographers have long used a variety of techniques to generate population estimates and projections. Population estimates for states, metropolitan statistical areas, and counties are prepared each year as a joint effort of the Census

Bureau and each state under the Federal-State Program for Local Population Estimates (FSCPE). The purpose of this program is to standardize data and procedures so that high quality estimates can be derived. Most states also generate population estimates and projections that are available through state agencies. For additional information on population projections and estimates.

Population estimates and projections generated by government agencies have historically been the only ones available. Today, however, a number of data vendors provide these figures. These vendor-generated data are often made available down to small units of geography (e.g., the census tract) and in greater detail (e.g., sex and age breakdowns) than government-produced figures. They may offer the flexibility to generate estimates and projections for "custom" geographies (e.g., a market area) not available for government-generated statistics. The drawback, of course, is that some precision is lost as one develops calculations for lower levels of geography and for population components. However, the ease of accessibility and timeliness of these vendor-generated figures have made them a mainstay of analysts and researchers in a number of disciplines.

### *Standardization*

Standardization is a method for adjusting mortality rates or other measures of vital processes for compositional factors that have an effect on those rates. For example, the number of deaths occurring in any year is a function of three components: health status, population size and demographic attributes (e.g., age). Since mortality rates are frequently used as indicators of health conditions, it is important to hold population size and age structure (and perhaps other attributes) constant when mortality rates are being constructed.

The calculation of rates addresses concerns over differences in population size and allows the analyst to compare the health status of two populations that are different demographically. The crude death rate (CDR), for example, may be used for this purpose. However, the CDR may be misleading since it is influenced by differences in the age structures of the populations in question. That is, areas with relatively young populations (and hence less risk of dying) are likely to report low death rates, while areas with relatively old populations (and greater risk of dying) are likely to report high death rates independent of the size of the respective populations. Thus, while age-specific death rates (ASDRs) are higher for all age cohorts in Mexico than in the United States, the CDR is lower in Mexico due to its very young age structure. For this reason, the unadjusted CDR is not a good measure for comparative purposes. It is possible to adjust or standardize rates in order to control for age structure and, often, other factors (e.g., race).

### *Cohort Analysis*

Cohort analysis is a method used by demographers to describe an aggregate of individuals having in common a significant event in their life histories, such as year of birth (birth cohort) or year of marriage (marriage cohort). The concept of cohort is useful because occurrence rates of various forms of behavior are often influenced by the length of time elapsed since the event defining the cohort—e.g., the rate of fertility is influenced by the event of marriage. The idea of cohorts is similar to that of generations, although some writers avoid the latter term because it also has other connotations.

Cohort analysis is used most extensively in demography on the assumption that tracking cohorts over time provides important insights into the cohort's behavior patterns. Life tables—a key mechanism employed by demographers to predict future populations—are based on cohort analysis.

### *Spatial Analysis*

Spatial analysis is a set of techniques for analyzing spatial data—that is data linked to a particular geographic point. Examining the spatial component of a demographic group extends traditional statistical analysis in order to investigate the geographic distribution of data. These techniques describe the distribution of data within geographic space (descriptive spatial statistics), analyze the spatial patterns of the data (spatial pattern analysis), identify and measure spatial relationships (spatial regression), and create a surface from sampled data (spatial interpolation). Software applications (geographic information systems) that implement spatial analysis techniques have become increasingly common.

***Spatial analysis allows demographers to link social groups and social phenomena to the environment.***

### *Epidemiologic Analysis*

Epidemiology is the study of patterns of disease occurrence and other health-related conditions within human populations and of the factors that influence these occurrences and conditions. Descriptive studies seek to demonstrate the association between various attributes while statistical analysis is used to explain relationships that are observed.

Ecologic studies—a form of epidemiologic analysis—are used to determine patterns of disease or disability across different populations, geographical areas and over time. Ecologic studies relate observed differences in morbidity or mortality to the spatial and temporal distribution of risk factors, lifestyles, genetic patterns and environmental exposure.

Analytic studies are used to test hypotheses related to epidemiological relationships. Analytic studies can be divided into two main types: experimental studies (clinical trials) and observational studies. In experimental studies the analyst seeks to control the variables. In contrast, an observational study typically requires that the analyst observe the phenomenon under study without attempting to influence it.

## Demographic Data

Data represents the fuel for the demographic engine and virtually all demographic analyses are data heavy. A demographer must be familiar with various types of data, their characteristics and their sources and be something of a statistician. The following sections introduce the student to this important component of the demographic endeavor. More detail is provided on demographic data in Chap. 6.

***Data represents the fuel for the demographic engine and virtually all demographic analyses are data heavy.***

### *Data Types*

Data utilized by demographers can take a variety of forms that include the following:

Raw data. Raw data represents the quantifiable attributes associated with any individual or group—e.g., the age of every student in a demography class. Each "record" (e.g., person, group) will have a series of attributes associated with it. Raw data by themselves are not very useful and need to be processed into "information" (as illustrated in the graphic below). Demographers conducting primary research generate raw data.

Statistics. Statistics are raw data that have been processed and subjected to some level of statistical analysis (that is, converted into information). Descriptive statistics provide a basic picture of the population (e.g., the age range and median age of the demography students) while inferential statistics examine relationships among various attributes of the population under study (e.g., the relationship between home state of students and political orientation).

Data sets. A data set is a collection of data in either raw or summary form. All of the attributes for all subjects in raw data form is one type of data set. The summarized data based on the raw data represents another type of data set (summary data). Data sets are typically utilized for analysis purposes when the demographer does not have access to the raw data.

Reports/Tables. Data presented in report or table form essentially represents a set of statistics. A table comparing the median of students in all classes in the university would be presented in table form. Most secondary data is presented in table or report format.

***Demographers convert raw data into information by processing, analyzing and interpreting it.***

Graphics/Maps. Graphics (charts, graphs) and maps represent methods for displaying data although they may serve as the basis for analysis. Charts and graphs typically depict summary data (e.g., the proportion of students of each age in the class) but may also depict the raw data (e.g., a scattergram of all ages). Maps illustrate the spatial (geographic) distribution of one or more attributes. Maps can typically display more information than charts and graphs and, by virtue of their link to geography, lend themselves to an examination of relationships between variables (e.g., the location of fast-food restaurants relative to the distribution of the low-income population). Box 2.1 describes the data conversion process.

**Box 2.1: The Data Conversion Process**

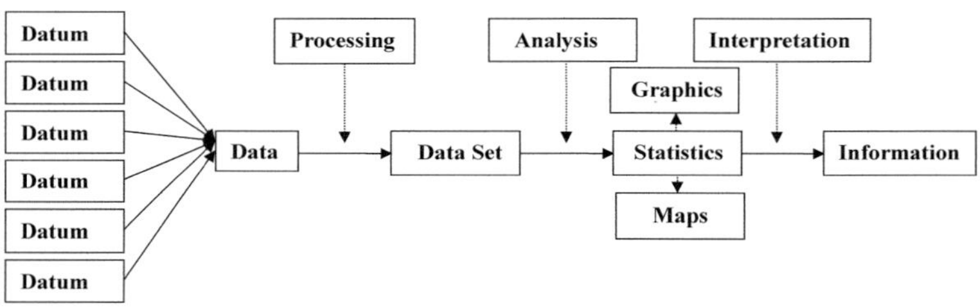


### *Data Sources*

Demographers acquire data for their analyses from a variety of sources. The primary sources of data are described below:

Censuses. A census involves a complete count of individuals (or entities) residing in a specific place at a specific time. The US Census Bureau (within the Department of Commerce) has conducted population censuses since 1790. Originally established as a means of determining the number of residents in each state for purposes of allocation of Congressional seats, the content of the census was expanded over time to include hundreds of additional data elements. Through the 2000 census, the census of population and housing collected basic data on each household and detailed data from a one-in-six sample. In 2010 the census collected only basic data on age, sex, race and household relationships. The American Community Survey (ACS) has been introduced by the Census Bureau as a replacement for the long-form data collection instrument for the decennial census.

Registries. A second method of data collection that generates information for demographers is represented by registration systems. A registration system involves the systematic compilation, recording, and reporting of a set of events, institutions, or individuals. The implied characteristics of a registry include the regular and timely recording of the phenomenon in question. Most registration systems relevant to this discussion are maintained by some branch of government, although other sponsors of registration systems exist as well.

The best-known registration activities in the United States are those related to "vital events", such as births, deaths, marriages, and divorces. The most extensive registration systems are maintained by the National Center for Health Statistics and the Centers for Disease Control and Prevention (CDC). Other useful systems are maintained by the Social Security Administration (SSA), the Center for Medicare and Medicaid Services (CMS), and Immigration and Customs Enforcement (ICE).

Sample surveys. A sample survey involves the administration of an interview form to a portion of a target population that has been systematically selected. The sample is designed so that the respondents are representative of the population being examined. This allows conclusions to be drawn for the total population based on the data collected from the sample.

The use of sample surveys has the advantage of being quicker and cheaper than a complete census and offers the opportunity to probe more deeply into the subject under study. Since surveys involve a sample, there is some slippage in accuracy relative to censuses. A serious shortcoming related to demography is the inability to compile adequate data through surveys for small geographic units due to limited sample sizes. Through various agencies, the federal government administers a number of ongoing surveys that involve information of interest to demographers.

***Demographic information is generated through a variety of means and is available from a wide range of sources.***

Synthetic data. Synthetic data refers to statistics produced in the form of estimates and projections. Synthetic data are generated by merging existing demographic data with assumptions about population change to produce estimates and projections. Estimates are generated for current or past time periods and projections for future time periods using a variety of different methodologies. Forecasts represent a variation on projections and are seldom generated by demographers.

### *Data Categories*

The following categories of data are most frequently utilized by demographers and others doing demographic analyses. The exhibit below presents some basic measures of central tendency.

Fertility data. Fertility data refers to information on the reproductive patterns of a population, including births and birth rates, characteristics of infants, mothers and fathers, sexual activities and contraception patterns, as well as reproductive health issues (e.g., infertility, premature births).

Morbidity data. One of the newer areas in applied demography, morbidity analysis examines data on the health status of the population and the types of health problems that it exhibits. Morbidity data is essential for conducting epidemiological analyses.

Mortality data. Mortality data refers to information on the deaths that occur within a population, including deaths and death rates, characteristics of the deceased, cause of death and factors that influence the mortality rate.

Migration data. Migration data refers to information on those migrating into, out of and within a particular geographic area. Data on internal migration depicts the movement patterns and characteristics of those moving within a particular country. Data on international migration depicts the movement and characteristics of those moving from one country (emigrants) to another country (immigrants).

Biosocial characteristics. Biosocial characteristics describe the basic demographic dimensions of a population's composition. Biosocial characteristics include age, sex, race and ethnicity—attributes that have both a biological basis and a significant social connotation. Biosocial attributes are present at birth.

Sociocultural characteristics. Sociocultural characteristics describe a population in terms of its social and/or cultural attributes and the relation of its members to the social structure. These include such attributes as marital status, income, education, occupation and religion. These represent acquired characteristics rather than traits ascribed at birth.

**Box 2.2: Basic Measures of Central Tendency**

Although demographers often make use of fairly sophisticated analytical techniques, there are some basic measures of central tendency that are useful as a starting point for demographic analysis.

Mean: The mean refers to the arithmetic average for an array of data points. This metric is calculated by summing the values for a particular attribute (e.g., income) and dividing the sum by the total number of cases. This is a useful metric for determining a group's average on a particular measure but suffers from the influence of outliers (i.e., abnormally high or low values) that may skew the average score.

Median: The median represents the mid-point value for an array of data points. This type of average determines the value at which half of the scores are above it and half are below it. This typically represents a more useful metric for demographers since it is not affected by outliers.

Mode: The mode is the score or interval that includes the greatest number of cases. In a normal distribution, the mode is typically somewhere near the middle since values normally cluster around the median.

## Key Points

- Demography has it own set of concepts, methods and data that set it apart from other disciplines.
- Demography provides a tool for dividing a population into meaningful subgroups.
- Demography provides observers with a means of understanding social phenomena.
- Knowledge of the demographic attributes of a population allows demographers to infer a lot about its other characteristics.
- Fertility, mortality and migration are the components of population change.
- Demographers use a variety of techniques for studying population characteristics and processes.
- Spatial analysis allows demographers to link social group and social phenomena to the environment.
- Demographic information is generated through a variety of means and is available from a wide range of sources.
- Demographers convert raw data into information by processing, analyzing and interpreting it.

## Discussion Questions

2.1 Demographers often generate projections for the size and characteristics of future populations. How does available demographic data allow demographers to project the size and characteristics of future populations?

2.2 If you were asked to do a planning study that would guide the development of a public school district, what types of demographic data would you need to acquire? In doing the study, what level of geography would be most useful for conducting the analysis (e.g., county, ZIP code, census tract)? Explain your choice of geographic level.

## Exercises

*Exercise 2.1: Converting Data into Information*

The basic demographics are provided for a typical college class. For each variable students are to calculate following statistics: mean, median and mode:

| Student | Age | GPA | Credit Hours |
|---|---|---|---|
| 1 | 18 | 3.0 | 15 |
| 2 | 18 | 2.5 | 15 |
| 3 | 19 | 2.5 | 30 |
| 4 | 18 | 2.2 | 12 |
| 5 | 20 | 3.5 | 24 |
| 6 | 20 | 3.0 | 15 |
| 7 | 18 | 2.0 | 12 |
| 8 | 17 | 3.0 | 15 |
| 9 | 22 | 3.0 | 45 |
| 10 | 20 | 2.1 | 24 |
| 11 | 18 | 2.5 | 15 |
| 12 | 20 | 2.0 | 30 |
| 13 | 18 | 2.5 | 15 |
| 14 | 19 | 2.5 | 15 |
| 15 | 20 | 4.0 | 30 |
| | Mean: ____ | Mean: ____ | Mean: ____ |
| | Median: ____ | Median: ____ | Median: ____ |
| | Mode: ____ | Mode: ____ | Mode: ____ |

*Exercise 2.2: Interpreting Data Tables*

The following table depicts trends in the size of the various age groups within a population over a period of 40 years. Convert the raw data into percents, examine the trends exhibited and draw conclusions about the future age distribution for this community

| Age Cohort | 1980 | | 1990 | | 2000 | | 2010 | |
|---|---|---|---|---|---|---|---|---|
| | No. | % | No. | % | No. | % | No. | % |
| 0-14 | 1,000 | ____ | 950 | ____ | 900 | ____ | 850 | ____ |
| 15-24 | 1,500 | ____ | 1,400 | ____ | 1,300 | ____ | 1,200 | ____ |
| 25-44 | 4,000 | ____ | 4,200 | ____ | 4,400 | ____ | 4,600 | ____ |
| 45-64 | 4,200 | ____ | 4,200 | ____ | 4,200 | ____ | 4,200 | ____ |
| 65-85 | 1,000 | ____ | 1,500 | ____ | 2,000 | ____ | 2,500 | ____ |
| 85+ | 100 | ____ | 200 | ____ | 300 | ____ | 400 | ____ |
| Total | ____ | | ____ | | ____ | | ____ | |

## References

Swanson, D. A., & Siegel, J. S. (2004). *The methods and materials of demography* (2nd ed.). Academic Press.

Thompson, W. R. (1929). Population. *American Journal of Sociology, 34*, 959–975.

## Additional Resources

Freyka, T. (2016). *The demographic transition revisited: a cohort perspective*. Max Planck Institute (Working Paper WP 2016–012). https://www.demogr.mpg.de/papers/working/wp-2016-012.pdf

McKeown, R. E. (2009). The epidemiologic transition: changing patterns of mortality and population dynamics. *American Journal of Lifestyle Medicine, 3*(1 Supplement), 195–265.

CHAPTER 3

# Population Size, Distribution and Concentration

**Abstract** The first "facts" of interest to demographers are the size, distribution and concentration of people within a geographic area. Methods for determining populatiion size are addressed and the influence of population size on demographic characteristics is discussed. Size is only one aspect of a population's description and information is provided on the importance of population distribution and how to measure it. The implications of population distribution for the characteristics of a particular geography are explored and trends in US population distribution reviewed.

## Introduction

The size of a population within a geographic area is the simplest and most straightforward of demographic attributes, and, for many purposes, it is the most important. Not only is the size of a population important in its own right but many of the calculations utilized by demographers and others depend on an accurate count of the population. After size, patterns of distribution and the concentration of persons within a geographic area are probably next in importance for anyone examining the interface of demographics and other aspects of society.

Demographers generally begin telling the "demographic story" about a population by presenting information on its size in very basic terms. Statistics that describe the size and location of various populations are, in fact, common in the popular press. Significant population increases (or decreases) often make the headlines and, when the census of population and housing is conducted every ten years, reports on population gains and losses are eagerly awaited. After all, a primary (and legally mandated) function of the census count is to



R. K. Thomas, *Demography: An Introduction to Population Studies*, Springer Texts in Social Sciences, https://doi.org/10.1007/978-3-031-56623-3_3

provide the basis for establishing Congressional districts. The significance of population size for the various states is demonstrated by the fact that, after the 2020 census, one state gained two Congressional seats and five states gained one additional seat. On the other hand, seven states lost one seat.

***Demographers generally begin telling the "demographic story" about a population by presenting information on its size.***

Data on the size and location of the population also provide the basis for calculating various rates for demographic attributes. Population counts serve as the denominators for statistics like the birth rate and death rate. Raw numbers may not have much meaning, so rates are calculated to facilitate comparisons between geographic areas and between populations of different sizes. The observation that deaths in California substantially exceed the number of deaths in Montana is not an indicator that health conditions are necessarily worse in California. A comparison of death rates—deaths per 1000 population, for example—would provide a better basis for a meaningful comparison. For many purposes, rate calculation is a critical step in determining a population's characteristics.

## Defining and Measuring Size, Distribution and Concentration

### *Population Size*

The size of a population is typically measured in terms of the number of individuals who reside within a defined geographic area at a specific point in time. The size of a population is an important consideration with regard to virtually every issue to be addressed. The fact that the United States has a population of around 335 million individuals has important national and international implications, as does the fact that Denmark has a population of less than 6 million residents (that is, smaller than 17 US states). The size of a population contributes to the level of diversity within the population and represents a key metric for demographers. The changing size of the US population continues to have numerous implications for the nature of our society. (Box 3.1 presents the US population for selected census years for the past 100 years.)

The question is often raised: does size matter? From a demographic perspective it does. As the size of a population increases (or decreases, for that matter), the dynamics of the group change. A shift from a small group to a large group, for example, has implications for attitudes and behaviors. The size of the population also determines the types of services it can support, from retail operations to healthcare. Every industry, in fact, has standards based on population size that determine the types of goods and services it can support.

***The size of a population determines the level of need for various services, contributes to the level of diversity within the population, and represents a key metric for demographers.***

Size is typically measured in terms of the number of individuals within a population in contemporary modern societies. However, this has not always been the case. In fact, when the first census was being planned in colonial America, there was serious discussion about including unlanded males, women, slaves and native Americans in the enumeration. (Unfortunately, there were no demographers around at that time to provide counsel on this topic.) More recently issues have arisen with regard to the counting of migrant workers and homeless people.

The individual is not the only unit that might be considered for an enumeration. More recently issues have arisen with regard to the counting of migrant workers and homeless people. Size of a population might be determined based on the number of households or families for example. In societies that emphasize communal life and de-emphasize the individual, they may see the world as made up of families rather than individual people. Other societies may see the household (which may or may not be comparable to a family) as the unit for measurement. While measuring size in terms of individual group members is technically the most accurate method, the perspectives of other cultures with regard to their "demographic map" should not be ignored.

**Box 3.1: Trends in US Population Size**

| *Year* | *Population* |
|---|---|
| 2020 | 331,449,281 |
| 2010 | 308,745,538 |
| 2000 | 281,421,906 |
| 1990 | 248,709,873 |
| 1980 | 226,545,805 |
| 1970 | 203,211,926 |
| 1960 | 179,323,175 |
| 1950 | 151,325,798 |
| 1940 | 132,165,129 |
| 1930 | 123,202,660 |

*Source* US Census Bureau

The most complete count of a population is performed by means of a census. A census involves an attempt to count (and describe) each individual "officially" residing in a country (or other geographic area) at a particular point in time. In the US, a census is conducted every ten years and, at the time of this printing, the final results from the 2020 census are being released.

***The most complete enumeration of a population is performed by means of a census that counts each individual residing in geographic area at a particular point in time.***

In order to assure a complete count of residents at the time of the census, the street address is used as the key locator. Using data provided by the US Postal Service, the Census Bureau identifies every postal address in the country and uses this information as the basis for developing its data collection methodology. For the majority of the population with a single permanent address, this approach is effective, and most census forms today are still returned by mail. However, there are circumstances when addresses may not exist that require additional effort on the part of the Census Bureau, such as in the case of migrant workers and homeless individuals. For those addresses for which no census form is returned, in-person interviews may be required.

In the process of gathering data on the number of persons in households and other living units, information on the unit's specific location (e.g., a street address) is collected as the geographic identifier for the person. The exception would be for individuals who live in group quarters such as a nursing home, college dormitory or prison, in which case the institution would be identified (although ultimately related to a physical address). Data collected at the household level are aggregated by the Census Bureau for various levels of geography (e.g., census tract, county, state). In addition, every population count must be specific with respect to date. Thus, data collected during the 2020 census are considered to be accurate as of April 1, 2020.[1]

Demographers and other users of census data are typically not interested in data on specific individuals and households (and, indeed, this information is not made available). Information on individuals is aggregated up to various levels of geography, with the smallest reported unit being the census block and the largest the United States as a whole. There are various levels of aggregation that occur between the smallest and largest units, and the level of aggregation most suitable for a particular analysis depends on the circumstances. The units of aggregation used by demographers are illustrated in Box 3.4.

***Census data on individuals are aggregated up to various levels of geography, with the smallest reported unit being the census block and the largest the United States as a whole.***

Several issues must be addressed in the counting of the population. Censuses are designed to count every individual regardless of their legal status. There has not been total agreement, however, as to who is a resident of the US and, thus, eligible to be counted. Today, with the increase in undocumented immigrants, new questions have emerged with regard to the counting of this

[1] The 2020 census was noteworthy for its high undercount rate and its incomplete coverage. (See Box 3.3.)

population. The Census Bureau has established rules that determine who is eligible for enumeration and clearly specifies who counts and who does not. So, for example, US military personnel and diplomats stationed abroad are counted as US residents while foreign tourists visiting in the US on census day are not.

Population counts are confounded in the US due to the high level of illegal immigration. Until the COVID-19 pandemic and Trump administration policies disrupted the steady flow of immigration to the US, approximately one million illegal immigrants were entering the US each year. The large number of undocumented residents represented a major challenge for census takers, especially when these immigrants do not want to call attention to themselves. Although individual information from the census is never released, many immigrants (even legal ones) are fearful to the intentions of the federal government and, thus, unwilling to participate in an enumeration.

***The enumeration of the US population is a challenge due to the high number of illegal immigrants.***

The assigned location of individuals in terms of geography is also of concern, with people generally counted at their usual place of residence. Since the place of residence is defined as the location where the individual usually sleeps and eats, the usual place of residence is easy to establish for the vast majority of the population. Definitional complications may arise, however, for subpopulations such as college students, migrant workers, persons with two or more residences, and the homeless. While college students are typically assigned to the location of their educational institution, the determination of usual residence for the remaining groups involves the application of specific rules, enumerator judgment, and special counting attempts (e.g., visiting rescue missions to count the homeless).

A further distinction between an individual's usual place of residence (nighttime) and his workplace location (daytime) is important. The downtown areas and industrial parks of many US cities teem with people during working hours only to become virtual ghost towns in the evening. The opposite is often true for residential areas. In many situations, information on the location of the daytime population may be more useful than information on place of residence. Locational decisions relating to medical emergency facilities, vehicles, and staffing, for example, require a sensitivity to differentials in daytime and nighttime populations. The ability to provide at least minimal services to all parts of the community requires an estimate of these distinct populations. While data on daytime populations are not routinely available, estimates of the daytime population for a geographic area can be estimated using such information as commuting patterns (e.g., from the population census) and commercial concentrations (e.g., from the economic census).

***Since 2005 the US decennial census has been supplemented by data collected through the annual American Community Survey.***

Since 2005 the US decennial census has been supplemented by the American Community Survey (ACS). With the 2010 administration, the census no longer administered its "long form", with the ACS intended to replace it. The American Community Survey uses a series of monthly samples to produce annually updated data for the same small areas (census tracts and block groups) formerly surveyed via the decennial census long-form. The ACS mails survey forms to over 3.5 million households annually. Data are collected primarily by mail, with telephone and personal follow-ups as required.

The ACS includes people living in both housing units and group quarters and collects data on the population related to age, sex, race and ethnicity, family and relationships, income and benefits, health insurance, education, veteran status, and disabilities. It also collects data on where people work and how they get there, where people live, and how much they pay for certain essentials. The ACS generates data for all levels of census geography except for the census block. The American Community Survey is considered an improvement over the census long form because it provides small-area information annually instead of once a decade.

**Box 3.2: The Changing Significance of the Decennial Census**

Since 1790 the decennial census has been considered the "gold standard" for collecting information on a population and has been emulated worldwide. Although its initial purpose was to serve as the basis for the determination of boundaries for federal Congressional districts, its usefulness has been expanded over the decades through the collection of detailed data (based on a sample) for the US population. Every ten years, the wide range of parties who use data for various purposes eagerly await the release of the new population statistics.

Over time, questions have been raised as to the benefits of the census, particularly in view of its cost and the growing complexity of its administration. Although every effort is made every ten years to assure a complete and accurate count, there is inevitably some slippage in data collection that results in an undercount of the population for that census year. Although statistical methods are utilized to address the issue of undercount, this problem has gotten worse as the US population has become more diverse and increasingly includes "undocumented" residents. This problem was particularly apparent with the 2020 census after the Trump administration attempted to interfere with the data collection process, and the COVID-19 pandemic corresponded with the time at which the census was being implemented.

The shortcomings of the census had long been recognized and the American Community Survey was instituted in 2005 to replace the Census Bureau's long form and allow for continuous annual updating of population counts. Although every effort was made to overcome the deficiencies of the 2020 census, demographers contend that this was the least accurate census in the history of the Bureau's data collection efforts. Many data users had already decreased their dependence on census data due to the small number of variables addressed and the lengthy time lag in producing the results (particularly at small levels of geography). Further, with the 2020 census additional measures were introduced ostensibly to assure confidentiality, thereby reducing the scope of the available data.

Since the decennial census is mandated by the US Constitution, there is no thought to eliminating this cumbersome and expensive initiative. However, it can be expected that users of population data will continue to pursue supplemental sources. In addition to the American Community Survey, other federal databases are being assessed as options for enhancing the available demographic data. Some commercial databases established by consumer marketing entities are thought to offer opportunities for supplemental data.

The future usefulness of the decennial census will no doubt depend on the type of support the Bureau's efforts receive, the Bureau's ability to assure an accurate count, policy decisions on the part of the Bureau, and the availability of alternative sources of data.

### *Population Distribution*

While the size of a population is an important consideration, the manner in which that population is distributed over a geographic area is equally important. It is one thing to know that the United States has a population of over 330 million residents, it is another to realize that these residents are unequally distributed throughout the country's various regions, states and communities. An important demographic dynamic for the US is constant "reshuffling" of its population among the various states.

The basic measure of population distribution relates to the *number* of people residing within various geographic areas. The US Census Bureau tracks the distribution of the population across the various states and other levels of geography. The nation's states are divided into regions and divisions for purposes of data collection and reporting. The nation's four major regions are comprised of nine divisions (as depicted in Box 3.3).

**Box 3.3: Census-Defined Regions and Divisions**

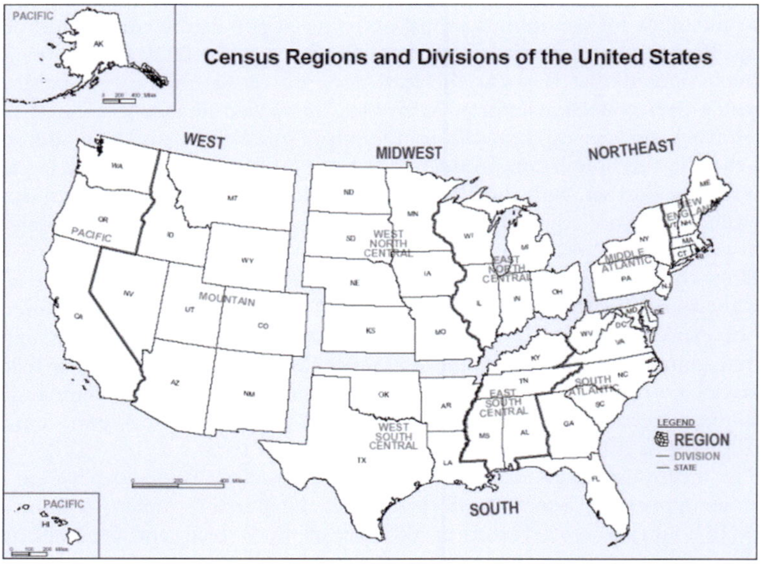


The number of residents reported for each state is useful information but for many purposes it may be more useful to think in terms of the proportion of the population accounted for by the respective states. While it is noteworthy that California contains nearly 40 million residents, this figure takes on more meaning when it is realized that these residents account for approximately 12% of the country's population. This is in stark contrast to the residents of the least populous state (Wyoming), with a population of less than 500,000 (or barely one one-hundredth of a percent).

Examining the population can be complicated enough when distribution of the total population is considered. Demographers, however, are typically interested in the distribution of subpopulations based on demographic attributes (e.g., race/ethnicity, income, age). The distributions of these subgroups are likely to differ in some cases significantly from the distribution of the general population.

Measures of population distribution depend on both accurate population counts and relatively stable geographic boundaries. Year-to-year state population estimates are not likely to be problematic since there are virtually never any changes in state boundaries. On the other hand, it is not unusual for the US Postal Service to modify ZIP Code boundaries in response to service demands. This means the number and proportion of residents allocated to a particular ZIP Code could change over time. Similarly, boundaries for metropolitan statistical areas expand through the periodic addition of counties, making it misleading to determine decade-to-decade change unless the boundaries are standardized. The boundaries for units of census geography are typically maintained from year to year and census to census, although there is always a certain amount of combining and disaggregating census geographies.

***A variety of different geographic units are used by demographers in their examination of population distribution.***

A variety of different geographic units are used by demographers in their examination of population distribution. These units may be as small as a census block or as large as a nation. They may be in easily recognized hierarchies that divide the nation into states and the states into counties or county equivalents. Other units for measuring may include less familiar units such as census block groups or county subdivisions. Some geographic units can be easily converted to others, while it may be difficult to reconcile the boundaries of some units with others (e.g., census tracts to ZIP Codes). In any case, an understanding of the geographic distribution of the population begins with an understanding of the various spatial units utilized as a framework for population distribution. Box 3.4 summarizes the various types of geographic units that might be of use to demographers and others using demographic data. Box 3.5 depicts the hierarchy of geographic units employed by the Census Bureau.

**Box 3.4: Units of Geography for Demography**

The geographical units used for demographic analysis can be divided into four major categories: political or administrative units, operational units, statistical units, and a residual category of units that do not fit into either of these categories.

*Political Units*

Political or administrative divisions are the most commonly used geographical units for most purposes. Many organizations' service areas coincide with political boundaries such as cities, counties, or states. Furthermore, it is convenient for private sector organizations to use standard political or administrative units to establish their boundaries.

Nation

A nation is defined by national boundaries. Data at the national level provides a context for the examination of population subdivisions within the framework of the total population. National averages (e.g., population growth rates, birth rates, mortality rates) are often important as a standard to which other levels of geography might be compared.

States

The major subnational political unit is the state, with data typically available for 50 states, the District of Columbia, and several US territories. Because the individual states have responsibility for a broad range of administrative functions, many useful types of data are compiled at the state level.

Counties

The county (or, in some areas, townships or parishes) represents the primary unit of local government. The nation is divided into more than 3100 county units (including some cities politically designated as counties). States typically report most of their statistics at the county level.

Cities

Cites are officially incorporated urban areas delineated by boundaries that may or may not coincide with other political boundaries. Although cities typically are contained within a particular county, many city boundaries extend across county lines. Because cities are incorporated in keeping with the laws of the particular state, the geographic unit does not make for a useful unit for comparative population analyses.

Congressional Districts

Congressional districts are established locally and approved by the federal government. These districts are typically delineated by means of political compromise and do not correspond well with any other geographical units. Although the Census Bureau reports its data for congressional districts, limited data are collected at the congressional district level.

State Legislative Districts

State legislative districts have similar characteristics to congressional districts. They are drawn up by the states based primarily on political compromise. Although the Census Bureau reports its data for state legislative districts, virtually no data are collected for such districts. Like Congressional districts, their boundaries are subject to periodic change.

*Operational Units*

Operational units are areas drawn up for the purpose of managing the operations of some entity. They may or may not have formal regulatory designation and, because they are responsive to the needs of a particular entity are more susceptible to change than some other types of units.

ZIP Codes

Unlike the geographical units previously discussed, ZIP Codes do not constitute formal government entities. Their boundaries are set by the US Postal Service and are subject to change as population shifts occur or as the needs of the Postal Service dictate. ZIP Codes seldom coincide with census tracts or other political or statistical boundaries, making the synthesis of data for various geographies extremely difficult. ZIP Codes tend to be much larger than census tracts, sometimes including tens of thousands of residents.

Utility Districts

Utility districts are designated for purposes of distributing power, water and gas and for disposing of garbage and other by-products of human activity. These districts tend to reflect the spheres of influence of various utility companies and may or may not correspond with other boundaries.

School Districts

School districts are established for the operation of school systems and are reflective of the educational needs of the local population. Although theoretically reflecting the distribution of school-aged children within the population, other factors may play a role in determining the configuration of school districts within a community. For most parts of the US continuous population shifts mean that the boundaries for school districts must be frequently adjusted.

*Statistical Units*

Statistical areas are established to allow various agencies of government to collect and report data in a useful and consistent manner. The guidelines for establishing most statistical units are promulgated by the federal government.

Regions

Regions are established for statistical purposes by the federal government by combining states into logical groupings. Four regions have been established by grouping states based on geographical proximity and economic and social homogeneity. Statistics are sometimes reported at this level by federal agencies.

Divisions

For statistical purposes the federal government divides the nation's four regions into nine divisions. Each division includes several states, providing a finer breakdown of the nation's geography. Divisions are seldom used as a basis for demographic analysis.

Metropolitan Statistical Areas

Metropolitan statistical areas (MSA) are delineated by the federal government as a means of standardizing the boundaries of cities and urbanized areas. Because each state has different criteria for the incorporation of cities, the MSA concept provides a mechanism for creating comparable statistical areas. An MSA includes a central city, central county, and any contiguous counties that could logically be included within the urbanized area.

Urbanized Areas

An urbanized area as defined by the Census Bureau includes the entire densely settled area in and around each large city, regardless of whether the area is within the corporate limits. Although limited amounts of data are available for urbanized areas, knowledge about urbanized areas is important in developing a full understanding of the population distribution within a metropolitan area.

Census Tracts

Census tracts are small statistical subdivisions of a county established by the Census Bureau for data collection purposes. In theory census tracts contain relatively homogeneous populations ranging in size from 1500 to 8000. For many purposes the census tract is the ideal unit for compiling demographic data. It is large enough to be a meaningful geographical unit and small enough to contribute to a fine-grained view of larger areas.

Census Block Groups

Census tracts are subdivided into census block groups that include approximately 1000 residents. A tract is composed of a number of block groups, each containing several blocks. The block group provides an even finer picture of a community than the tract level, although fewer data elements are likely to be compiled at the block group level.

Census Blocks

Census block groups are subdivided into census blocks, the smallest unit of census geography. The term *block* comes from the fact that the typical block is bounded on four sides by streets, although some other visible feature may serve as a boundary. Census blocks tend to be the most homogeneous of any unit of census geography, with the average block housing approximately 30 persons. Only a limited amount of demographic data are available for census blocks.

ZIP Code Tabulation Areas

ZCTAs are generalized area representations of US Postal Service ZIP Code service areas. They are created by aggregating the blocks whose addresses use a given ZIP Code into a ZCTA with that ZIP Code assigned as its ZCTA code. ZCTAs represent the majority of US Postal Service five-digit ZIP Codes found in a given area. The Census Bureau's intent was to create ZIP Code–like areas that would retain more stability from census to census.

*Other Units of Geography*

Natural Regions

There are certain geographic regions that are distinguished enough in terms of some physical, social or economic trait to be recognized as a distinct area. Some such areas may have been historically recognized as unique geographic areas and these may occasionally be formally labeled as such for some practical purpose. Examples that come to mind include Appalachia and the Mississippi Delta. The federal government has formally recognized these to natural regions through the establishment of federal agencies (regional commissions) for addressing problems facing the respective regions.

Areas of Dominant Influence

Taken from media advertising, the *area of dominant influence*, or ADI, refers to the geographic territory (typically a group of counties) over which a form of media (e.g., television, newspaper) maintains predominance. This concept is useful when healthcare marketers are interested in media promotions and want to determine the reach of a particular marketing campaign.

**Box 3.5: Overview of Census Geography**

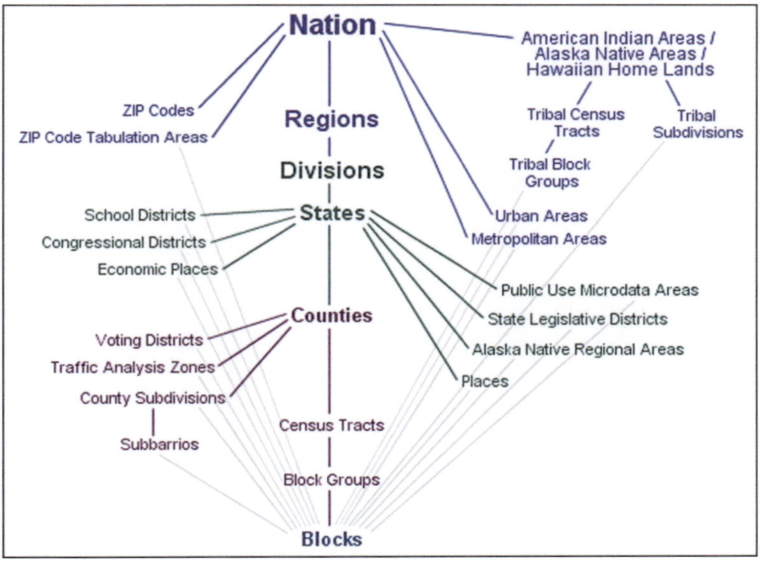


## *Population Concentration*

The extent to which a population is concentrated represents a form of distribution and relates primarily to how clustered or dispersed a population is. Concentration is usually measured in terms of density, and this specialized measure of distribution provides information on a population's concentration within a particular geographic area. Usually measured in the US in terms of persons per square mile, density measures indicate the extent to which the population is concentrated or dispersed within a particular geographic area. In other countries, density may be measured in terms of square kilometers, hectares, acres or some other measure. Density can be computed for any area—as long as one knows the size of the land area and the population within that area. Note that land area is emphasized since bodies of water are typically excluded when calculating density. Similarly, any areas where inhabitation is restricted (e.g., national parks) may also be excluded from the denominator in density calculations.

***Measures of population density indicate the extent to which the population is concentrated or dispersed within a particular geographic area.***

The population density of cities, states, entire continents, and even the world can be easily computed. Monaco, with an area of less than one square mile, has the world's highest population density or almost 43,000 people per square mile. Mongolia is the world's least densely populated country with only 4.3 people per square mile. Australia is a close second with 6.4 people per square mile. The United States, with its extensive territory, has a current population density of around 84 persons per square mile.

Another way of looking at concentration is in terms of an area's mean center of population. This is the point at which the population is equally distributed in all directions at that point so that the area would balance like a plate set on top of a stick if all persons were standing on the "plate." At the time of the first census in 1790 the mean center of population for the US was in Maryland; today it is in Missouri. During the twentieth century the mean center of population shifted 324 miles to the west and 101 miles to the south.

Obviously, in the US and in much of the world for that matter, the degree of population concentration has been increasing. The estimated population density for the US in 2020 was around 94 persons per square mile (i.e., 331 million residents divided by 3.8 million square miles). The most highly concentrated state population is found in New Jersey, with 1258 persons per square mile in 2020; the least highly concentrated state population is found in Alaska, with barely one person per square mile. The borough of Manhattan in New York City is considered the highest concentration of population within the US, with approximately 29,000 persons per square mile. Box 3.6 presents population density by state for 2020.

Like distribution, measures of concentration assume reasonably accurate population counts and fixed geographic boundaries. For most units of geography (e.g., states, counties, census tracts) the boundaries and, hence, the land area typically do not change over time. For other units such as ZIP Codes and school districts the boundaries are subject to change.

**Box 3.6: Population Density United States 2020**

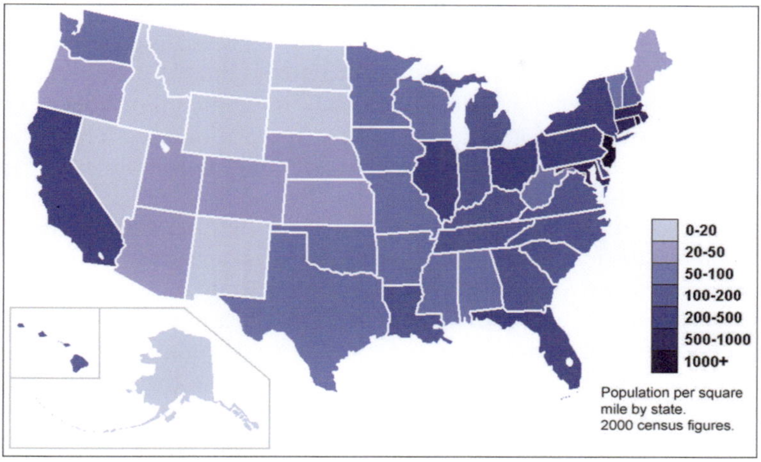


*Source* US Census Bureau

## Community Type

Demographers classify populations in terms of the types of communities in which they reside. While the type of community is addressed somewhat by the geographic units discussed above, there are other less formal community types that demographers may employ in their analyses. These not only describe the nature of population distribution, but they reflect differences in lifestyles and various demographic characteristics.

Demographers classify the most sparsely populated areas as rural areas. From the Census Bureau perspective these are areas containing less than 2500 residents. The Bureau further divides rural areas into rural farm areas and rural non-farm areas. With this low threshold for being considered rural, many states remain predominantly rural today.

***Demographers classify populations in terms of the types of communities in which they reside.***

Generally speaking, communities of 2500 or more residents (with at least a certain population density) and less than 50,000 residents are considered "small towns" or just "towns" in popular parlance. It is usual to have small towns scattered around rural areas; however, increasingly, the suburbs and exurbs of metropolitan areas contain small towns, either spontaneiously generated (new towns) or incorporated into the larger metropolitan area through annexation (old towns).

Typically, communities of 50,000 or more are considered to be "cities," with social, cultural and economic characteristics that separate them from towns. Cities are further classified as small cities, medium cities and large cities, with essentially no upper limit on the population size. The thresholds for different sizes of cities are not based on scientific criteria but are a function of who is doing the classifying.

Cities of any significant size are classified by the Census Bureau as metropolitan statistical areas, micropolitan statistical areas, or combined statistical areas. The first two invariably contain a central city of 50,000 or more for a metropolitan area or a city of 10,000 or more for a micropolitan area. With rare exceptions, these urban areas contain a central county which houses the central city, along with adjacent counties that comprise the metro- or micro-area. These adjacent counties are typically referred to as suburban counties and the urban communities within their borders as suburbs. Further, the central county often includes suburbs that are distinct from the central city. "Exurbs" are snall towns located outside MSA boundaries that have some relationship with the metro area. Combined statistical areas (CSAs) are metropolitan or micropolitan statistical areas that retain their metro- or micro- identity while being included in a larger metropolitan area.

In certain parts of the country, most notably the East Coast and West Coast, there are unbroken stretches of urban development in which metropolitan areas actually merge with each other. These are referred to as consolidated metropolitan statistical areas and may contain tens of million residents.

***In certain parts of the country there are unbroken stretches of urban development in which metropolitan areas actually merge with each other.***

The central city is also often subdivided into an urban core (that typically includes a central business district) and surrounding suburban areas (which may or may not be within the city limits). Older cities may have two or more successive rings of suburban development, often encouraged by the establishment of highway "loops."

Finally, communities and counties that exist beyond the suburban counties of a metro- or micro-area are referred to as exurbs. These may include rural areas, small towns and/or cities that are somewhat beyond the Census Bureau-recognized limits for consideration as part of a metropolitan or micropolitan area.

Understanding these various gradations of both official and unofficial community types would benefit from a concrete example. Officially part of the Washington-Arlington-Alexandria, D.C., Virgina, Maryland and West Virginia Metropolitan Statistical Area, the Washington, D.C., area illustrates the various community types. The District of Columbia, while not techically a county, essentially has the status of the central county in the MSA. Washington is the central city in that "county" which contains other, smaller cities (e.g., Georgetown). This central "county" (District of Columbia) is surrounded by counties in three states, Maryland, Virgina, and West Virginia. These "suburban" counties include cities of various sizes (e.g., Arlington, Virginia, and Silver Springs, Maryland), towns of various sizes (some incorporated, others unicorporated) and rural areas (including some rural farm areas). There are an additional eight counties in Maryland and Virginia that are considered exurban (i.e., not part of the metropolitan area but nevertheless interconnected). Finally, the Washington metropolitan statistical area is combined with the Baltimore Metropolitan Statistical Area to form the Washington-Baltimore Consolidated Statistical Area. (See Box 3.7 for a map showing certain aspects of the Washington MSA.)

**Box 3.7: Washington, DC, Metropolitan Area**

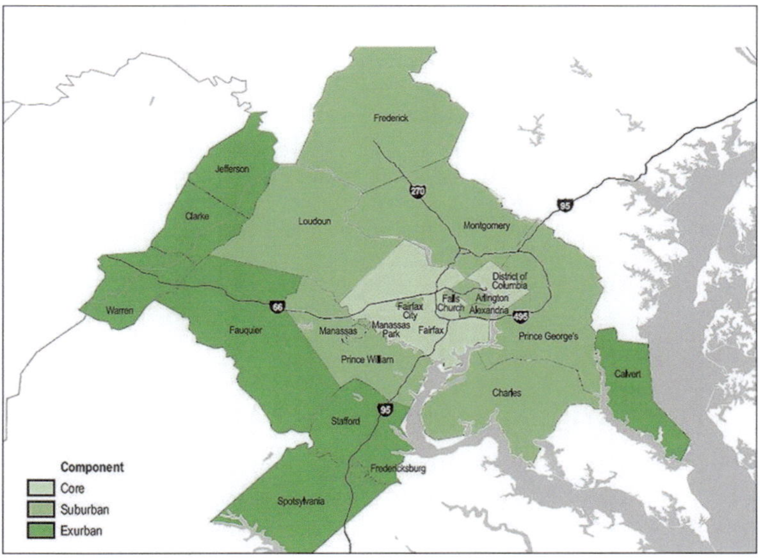


This chapter has illustrated the breadth of data available for examining population size, distribution and concentration. With such a wide range of data options, it may be difficult to determine which level of geography best suits the aims of an analyst. Box 3.8 provides some guidance in determining the degree of granularity appropriate for a particular demographic analysis.

**Box 3.8: Choosing the Appropriate Geographic Level**
A number of different "geographies" are available to demographers undertaking a demographic analysis. The choice of geographic level to be used depends on: (1) the objectives of the analysis; (2) the degree of granularity required; (3) the implications of crossing political boundaries; and (4) the type and quality of the data. The last criterion is worth expanding upon since data of different types and degrees of granularity vary with geographic level. At the national level, for example, demographic data can be obtained in great detail for virtually any desired variable. These data are readily available and can be disaggregated for any relevant demographic category (e.g., Hispanics, frail elderly, baby boomers).

Similarly, at the state level a wide range of data is available and, in some cases, exceeds in scope what data may be available at the national level due to state-specific programs. At the county level, a considerable amount of data is available, with counties being the initial source of data on vital statistics (e.g., births, deaths). Data collected through the decennial census are available for all of these levels, although this information quickly becomes dated. (See the discussion on the American Community Survey in this regard.)

At the census tract level and below, data are generated through the decennial census and the American Community Survey. As noted above, there is extensive data available at the tract level, somewhat less data available at the block group level, and only limited data available at the block level. Although the Census Bureau does allocate census data to the ZIP Code level, the best source of data at this level is the commercial data vendors who specialize in generating estimates and projections for ZIP Codes.

Ultimately, every situation will require a careful assessment of the data that would be desirable, the data essential to have, and, significantly, the data likely to be available at the appropriate geographic level.

## Population Change and Its Measurement

Demographers pay considerable attention to population change as this reflects the dynamic aspects of an area's demographics. A population's characteristics at any point in time represents a static "snapshot" of that population, and it is the changes that occur in a population that represent the dynamic aspects. Population change can be manifested in a change in the size of a population, in the distribution and concentration of the population, or in the population's composition.

There are three components to population change to be considered: fertility, mortality and migration. Individuals are added to the population through the fertility process and subtracted through the mortality process. Individuals are added to a population through in-migration and subtracted through out-migration. Thus, the population at Time 2 is a result of adding births and in-migrants and subtracting deaths and out-migrants from the population at Time 1. Population change resulting from the addition of births and the subtraction of deaths is referred to as "natural increase" (or decrease as the case may be). In a closed society, these are the only factors that influence population change. Since there are virtually no closed societies, the role of migration must be factored into the equation. (Each of the components of population change will be discussed in greater detail in subsequent chapters with more detailed information provided on the process of population change.)

***There are three components to population change to be considered: fertility, mortality and migration.***

Changes in population size and distribution almost invariably result in changes in population composition. Even the most static traditional society will undergo change as a result of demographic processes, albeit at an almost imperceptible rate. Differential birth rates and death rates for subgroups within society may result in compositional change. This is occurring in the United States today, for example, as the white population exhibits lower birth rates than most minority populations, leading to a restructuring of the racial and ethnic composition of the US. population. Similarly, in-migrants may have different characteristics from the population in the receiving community with migrants likely to affect the population composition of both the communities from which they originate and those to which they migrate.

While the measurement of population change is a critical activity of demographers, there are often situations in which adequate information is not available for this purpose. There are often occasions in which information on population size and characteristics is desired for a past time period for which data are not available or for a future time period for which data are obviously not available. In these situations, demographers rely on population estimates and projections to determine population change.

***Changes in population size and distribution almost invariably result in changes in population composition.***

Population estimates and projections can be simple or complex. The simplest approach to estimating or projecting change is to use known data points to extrapolate or interpolate data. Obviously, the more data points one has access to, the more accurate the estimate or projection. If one knows, for example the population of a community for 2010 and 2020, respectively, the population estimate for 2015 can be generated by interpolating from these two figures. Assumptions must be made, of course, concerning the rate of change for that decade, with the most straightforward approach assuming that the rate of change is constant over the 10-year period.

Similarly, if one wanted to project the population for 2030 for this same community, it would be possible to extrapolate from the existing data points into the future. The analyst could determine the numerical change between 2010 and 2020 and assume that same absolute change for 2030. Alternatively, the analyst could determine the percentage change and apply that figure and assume the same proportional change for 2030. In most cases, these two approaches yield different answers, and the demographer must have an understanding of the community in question to determine the best method to use.

Estimates and projections can also be produced by combining data on the components of population change (i.e., births, deaths, and migration) and/ or from "symptomatic data" (e.g., housing starts and utility hookups. Reality-based methods such as these are preferred if adequate data are available.

Population estimates and projections are viewed as proxies for actual data such as census data. They are, however, susceptible to statistical error and the amount of error may not be known. Further, their efficacy depends on the accuracy of the assumptions made. (See Chap. 10 for more detail on population estimates and projections.)

## Trends in Population Size, Distribution and Concentration

Since the first US census was conducted in 1790 (and before) the population of the United States has experienced steady growth. The continuous increase in the size of the population has been a major force in the political and economic development for the country. From a nation of roughly 4 million people in 1790, the population of the United States increased to over 330 million in 2020. The increase over each of the ten-year periods has demonstrated some fluctuation, with the highest decadal growth rate (18.5%) during the 1950s. While the high growth of the 1950s reflects the high post-war fertility rates that spawned the baby boomers, the spike in growth during the 1990s primarily reflects an increase in the volume of immigration. The growth trend for the United States from 1950 to 2020 is presented in Box 3.9. (See Chap. 11 for additional information on demographic trends in the US.)

**Box 3.9: Population Growth and Density United States 1950–2020**

| *Year* | *Population* | *Population per square mile* | *Increase over previous decade* | *Percent change* |
|---|---|---|---|---|
| 2020 | 331,449,281 | 93.8 | 22,703,743 | 7.4 |
| 2010 | 308,745,538 | 81.5 | 27,323,363 | 9.7 |
| 2000 | 281,421,906 | 79.5 | 32,703,605 | 13.1 |
| 1990 | 248,718,301 | 70.3 | 22,176,102 | 9.8 |
| 1980 | 226,545,805 | 64.0 | 23,243,774 | 11.4 |
| 1970 | 203,302,031 | 57.4 | 23,978,856 | 13.4 |
| 1960 | 179,323,175 | 50.6 | 27,997,377 | 18.5 |
| 1950 | 151,325,798 | 42.6 | 19,161.229 | 14.5 |

*Source* US Census Bureau

Population density in the United States is currently estimated at around 94 persons per square mile. While density has been increasing steadily over time, these figures are relatively low compared to most other countries. The 2020 US figure of 83.9 per square mile compares to the most recent estimate of 398 for China, 623 for Germany, 1214 for India, 895 for Japan, and 995 for Belgium, just to provide a few examples. Population density at the national level is of limited interest to demographers because of the small amount of decade-to-decade change. However, when subnational geopolitical units such as regions, states, counties and cities are considered, the amount of change in the concentration of persons is typically much greater.

An examination of growth rates for the various states and the District of Columbia provides insight into not only population growth and change but into population redistribution. Box 3.11 presents data on the states with the highest and lowest growth rates for the 2000–2010 and 2010–2020 time periods.

**Box 3.10: Population for the 10 States with the Highest and Lowest Growth Rates 2000–2010 and 2010–2020**

| *2000–2010* | | | *2010–2020* | | |
|---|---|---|---|---|---|
| *Rank* | *State* | *Percentage Change* | *Rank* | *State* | *Percentage Change* |
| 1 | Nevada | 35.3 | 1 | Utah | 18.4 |
| 2 | Arizona | 24.7 | 2 | Idaho | 17.3 |
| 3 | Utah | 23.9 | 3 | Texas | 15.9 |
| 4 | Idaho | 21.2 | 4 | North Dakota | 15.8 |

| | | | | | |
|---|---|---|---|---|---|
| | Texas | 20.9 | 5 | Nevada | 15.0 |
| 6 | North Carolina | 18.6 | 6 | Colorado | 14.8 |
| 7 | Georgia | 18.5 | 7 | District of Columbia | 14.6 |
| 7 | Florida | 17.9 | 8 | Florida | 14.6 |
| 9 | Colorado | 17.0 | 9 | Washington | 14.6 |
| 10 | Texas | 10.2 | 10 | Arizona | 11.9 |
| 51 | Michigan | 0.0 | 51 | West Virginia | – 3.2 |
| 50 | Vermont | 0.0 | 50 | Mississippi | – 0.2 |
| 49 | New York | 0.2 | 49 | Illinois | – 0.1 |
| 48 | Ohio | 0.2 | 48 | Connecticut | 0.9 |
| 47 | West Virginia | 0.3 | 47 | Michigan | 2.0 |
| 46 | Massachusetts | 0.3 | 46 | Wyoming | 2.3 |
| 45 | Maine | 0.4 | 45 | Ohio | 2.3 |
| 44 | Pennsylvania | 0.4 | 46 | Maine | 2.6 |
| 43 | Rhode Island | 0.4 | 43 | Louisiana | 2.7 |
| 42 | Iowa | 0.4 | 42 | Missouri | 2.8 |

*Source* US Census Bureau

A major trend in US population distribution has been the steady migration westward. Beginning with the settlement of the original colonies on the east coast, there has been fairly constant population movement from east to west. The decennial census represents a useful source of data to track this movement. Box 3.11 depicts the population center for the US for every census from 1790 to 2020. This population center is calculated as if the nation were a plate that everyone stood on and the point a which the plate would balance would be considered the population center. In 2020, US population center was somewhere in south-central Missouri.

There are several trends related to community type that should be noted within the context of overall mobility patterns. Historically, the US has been considered a population of movers, with much of the population changing residence each year. In the twenty-first century, the nation has experienced a decrease in the frequency of residential moves, with a mobility rate that would be even lower were it not for the large number of immigrants entering the nation and subsequently relocating throughout the country.

## Box 3.11: Trends in the US Population Center

Mean Center of Population for the United States: 1790 to 2020

Within this context of declining overall mobility, there are several noteworthy trends that have played out over the past decades. The first major trend was the urbanization of the population around the beginning of the twentieth century. This trend was driven by an increasingly industrial society that essentially converted the US population from an agrarian one to an urban one. By 1920, a majority of Americans lived in cities, and this trend continued steadily through World War II. During the post-war period, population movement shifted from the central city within urban areas to outlying residential areas, with the suburb becoming the predominant form of residence by the 1970s. A sub-trend occurred late in the twentieth century as suburbs became increasingly inhabited by minority groups that were previously concentrated in the inner city.

By the beginning of the twenty-first century the movement to the suburbs had diminished and two other albeit less dominant trends emerged. One of these was the exurban movement as urban and suburban dwellers moved further away from the central city to distant "exurbs." The other trend was embodied by the "gentrification" movement that saw a return flow of residents to older urban areas. This trend usually involved efforts to revitalize older sections of cities, often displacing long-time residents. Much of the movement back to the central city was driven by Millennials who had limited interest in homeownership and frequently did not own automobiles.

The most recent trend in residential mobility was spawned by the movement toward remote work arrangements. Remote workers can live almost anywhere, and this is beginning to have an impact on the nation's population distribution. The trend toward remote work was further exacerbated by the COVID-19 pandemic that drove workers from their traditional brick-and-mortar workplaces. While the long-term implications of remote work for population distribution have yet to be determined, it does appear that a definite (and permanent) shift has taken place. This has been most felt in expensive, high-density urban areas that have experienced an outflux of remote workers. Smaller urban areas, vacation locales and areas that offer cultural amenities have been the beneficiaries of this trend (Economic Innovation Group, 2022).

***Much of this movement back to the central city has been driven by millennials who have limited interest in homeownership.***

## Sources of Data on Population Size, Distribution and Concentration

The decennial census in the United States has historically been the primary source of data for determining population size, distribution and concentration. A census is never going to count every single resident and the US census is no exception. Censuses in the US typically exhibit a 2–3% undercount of residents. A challenge for the implementation of the census is the growing number of undocumented aliens who often have an incentive to not be enumerated. Because of the inevitable undercount, there is controversy surrounding every census and the discussion of the undercount has become increasingly politicized. Obviously, the stakes are high and different constituencies line up on either side of whether to accept the official (but admittedly flawed) count from the Census Bureau or to adjust the count to take into consideration the undercount. Since Congressional seats are allocated based on the final census count, states that benefit from an overcount tend to support the official tally while those who suffer from an undercount tend to support an adjusted figure.

The picture is complicated by the fact that the undercount is not spread randomly across the population. Certain hard-to-reach groups such as minorities, poverty populations, migrant workers, immigrants and the homeless tend to be undercounted. The fact that these groups exhibit greater or lesser concentrations in different parts of the county means that the impact of the undercount is highly variable. Because of these issues, the Census Bureau expanded its efforts to reach these populations and thus obtain a more accurate enumeration starting with the 2010 census.

Today, the decennial census is supplemented by the American Community Survey (ACS), and the ACS has already replaced the census long-form administered prior to 2010. The American Community Survey uses a series of monthly samples to produce annually updated data for the same small areas (census tracts and block groups) formerly surveyed via the decennial census long-form. Data are collected primarily by mail, with telephone and personal follow-ups as required. The ACS includes people living in both housing units and group quarters.

The ACS collects data on the population related to age, sex, race and ethnicity, family and relationships, income and benefits, health insurance, education, veteran status, and disabilities. It also collects data on where people work and how they get there and where people live and how much they pay for certain essentials.

It is worth noting that there is another non-government source of data on the US population, and one that the Census Bureau is investigating as a potential adjunct to the ACS. This involves the national consumer databases compiled by various business and marketing enterprises. These databases are generally used for marketing purposes and not deliberately designed for demographic data collection. Nevertheless, these national databases are being increasingly considered as supplements to the decennial census.

## Key Points

- Demographers generally begin telling the "demographic story" about a population by presenting information on its size.
- The size of a population determines the level of need for various services, contributes to the level of diversity within the population, and represents a key metric for demographers.
- The most complete enumeration of a population is performed by means of a census that counts each individual residing in geographic area at a particular point in time.
- Census data on individuals are aggregated up to various levels of geography, with the smallest reported unit being the census block and the largest the United States as a whole.
- Since 2005 the US decennial census has been supplemented by data collected through the annual American Community Survey.
- A variety of different geographic units are used by demographers in their examination of population distribution.
- Measures of population density indicate the extent to which the population is concentrated or dispersed within a particular geographic area.
- Demographers classify populations in terms of the types of communities in which they reside.
- Changes in population size and distribution almost invariably result in changes in population composition.
- In certain parts of the country there are unbroken stretches of urban development in which metropolitan areas actually merge with each other.
- The three components of population change to be considered are fertility, mortality and migration.

## Discussion Questions

Since most of you in this class are young adults, what factors do you consider when determining (1) where you would like to live and (2) where you would like to work?

What differences can be observed between the characteristics of rural areas, small towns and major cities? How does size affect the social structure and cultures of communities of various sizes?

To what extent do residents group together in terms of age? To what extent can communities dominated by specific age cohorts be identified? What are the social and cultural implications of the establishment of communities based on age?

## Exercises

*Exercise 3.1: Geographic Unit Identification*

For this exercise, students will access the Census Bureau website in order to identify various units of geography with which they should be familiar. Follow the steps below and identify the geographic units requested relevant to your residence.

(1) Access the Census Bureau website at https://geocoding.geo.census.gov/geocoder/geographies/address?form.
(2) Complete the information requested on the form (down through Zip Code)
(3) Click on "Get Results." A list of geographies associated with that address will be generated.
(4) Search the list to find the following descriptors for your address:

   a. Census tract (Tract Code): _______
   b. Census block (Block Code): _______
   c. Census block group (GEOID): ______

(5) Refer to Box 3.5 to obtain the following information:

   a. Region
   b. Division

(6) Now return to the computer to access a census map:

   https://www.census.gov/geographies/reference-maps/2020/geo/2020-census-block-maps.html
(7) On the page displayed in the "County or County Equivalent" box, select your state from the pull-down menu.
(8) On the list that is displayed, select your county.
(9) Click on the.pdf link (it will be different for each county)
(10) Locate the general area of your residence on the map.

(11) Locate the grid number listed for your area in the list of maps displayed to the left.
(12) Open the selected map and locate your residence. (You may have to zoom in using the + sign at the top.)
(13) When you find the block you residence is located on verify the following:

a. Census tract (large number)
b. Census block (small number)

## Reference

Economic Innovation Group. (2022). *How remote work is shifting population growth across the US*. https://eig.org/how-remote-work-is-shifting-population-growth-across-the-u-s/

### Additional Resources

Baker, J., Swanson, D. A., Tayman, J., et al. (2017). *Cohort change ratios and their applications*. Springer.
McCullough, M. (2023). Policy map: Easy on-line mapping for business users. *CIO Review*. https://www.cioreview.com/policymap
Momoh, O., Murry, C., & Rathburn, P. (2022). *Population definition in statistics and how to measure it*. investopedia. https://www.investopedia.com/terms/p/population.asp
OpenStax. (2021). *Introduction to sociology*. https://openstax.org/details/books/introduction-sociology-3e
Smith, S. K., Tayman, J., & Swanson, D. (2014). *A practitioner's guide to state and local population projections*. Springer.
Thompson, W. S. (1929). Population. *American Journal of Sociology, 34*(6), 959–975.

CHAPTER 4

# Population Composition: Biosocial Characteristics

**Abstract** This chapter gets at the "heart" of the demographic endeavor as it describes the various attributes that are used to describe populations from a demographic perspective. This chapter considers the biosocial characteristics that define a population and describes the manner in which characteristics like age, sex and race not only serve to profile a population but have important implications for the demographic processes. Trends in the composition of the US population are reviewed in this chapter and the changing nature of US society described.

## Introduction

*Population composition* refers to the combined demographic characteristics of persons within a geographic area. These characteristics create a profile for a population and are the attributes that give a population its particular character. The population composition of New York City sets it apart from Memphis, Tennessee, and Omaha, Nebraska, more so than its size. Compositional variables are primarily descriptive in nature. Their usefulness is derived from their ability to profile a population in terms of its relevant attributes. An area's age distribution, racial makeup, income level, and dominant religion are the types of characteristics that give a population its "personality."

## Compositional Variables

Compositional variables can be divided into two categories: biosocial characteristics and sociocultural characteristics. *Biosocial characteristics* are those that have an underlying biological or physical component. As such, they tend





R. K. Thomas, *Demography: An Introduction to Population Studies*, Springer Texts in Social Sciences, https://doi.org/10.1007/978-3-031-56623-3_4

to be "ascribed" characteristics present at birth and not amenable to change. Biosocial factors include age, sex, race and ethnicity, and with the exception of ethnicity all are rooted squarely in biology. Ethnicity has its basis in a common cultural heritage, but endogenous marriage within ethnic groups often results in the development of a gene pool that fosters common physical characteristics.

Biosocial characteristics have significant social connotations in that society ascribes certain attributes to them. For example, while being male or female is a biological condition, each society attaches certain attributes to men (i.e., masculine traits) and to women (i.e., feminine traits). In fact, demographers use the term "sex" to refer to biological differences and "gender" to refer to socially ascribed attributes.

***Compositional variables can be divided into two categories: biosocial characteristics and sociocultural characteristics.***

*Sociocultural factors* reflect the position of society members within the social structure. Sociocultural factors, in US society at least, are primarily "achieved" rather than ascribed. These are not traits one is born with in a biological sense but those that are acquired (voluntarily or involuntarily) through one's place in the social system. These factors are "cultural" in that those affected take on characteristics assigned by society. Sociocultural factors include marital status, income, education, occupation, and religion among others.

This chapter focuses on biosocial characteristics, with sociocultural characteristics addressed in Chap. 5.

### *Age Distribution*

For many purposes, the age structure of a population represents the most significant compositional variable. After population size, the age distribution is the most important factor in determining a society's character and for calculating many of the rates used by demographers. Age is measured in chronological terms beginning at a person's date of birth. Age data in the United States are generally thought to be of high quality, although in survey research some age "heaping" is seen in self-reported data for milestone years such as 21, 62, 65, 100, and years that end in zero.

***For many purposes, the age structure of a population represents the most significant compositional variable.***

Demographers sometimes speak of the age-range of a population. While this is not very meaningful when considering the nation, a state or even a community, if you are considering a relatively circumscribed group such as a college classroom, a social organization or a workforce this may be a useful concept. Consider, for example, the implications of a college class with an age range of 18–21 compared to a class with an age range of 18–50. As with other metrics related to age distribution, there are times when it is useful to determine the median or mean age of the class or other group as well as the modal age category.

Although age data may be sometimes presented in single years (e.g., 1, 2, 3, etc.), ages for a population are typically grouped into intervals to simplify data presentation. Five-year and 10-year intervals are generally used by demographers, with exceptions sometimes made for the youngest intervals (under 1, 1 to 4), and the oldest intervals (age 85 and above). However, there is no substantive reason for utilizing these intervals, and others that are more relevant to the issue at hand may be appropriate. At times, it may be worthwhile to create more "functional" intervals, such as 0–14, 15–24, 25–44, 45–64 and 65 and older. This method generates logical groupings that reflect relevant stages in the life cycle.

Measures of statistical dispersion (e.g., standard deviation) may also be used to describe a population's age distribution. Box 4.1 contains detailed age data for the US population in 2021. The data show that the distribution is a mature one with similar proportions at all age groups except the very oldest.

**Box 4.1: Age Distribution of US Population: 2021 (in Thousands)**

| *Age category* | *Population* | *Percent* |
|---|---|---|
| Under 5 years | 18,661 | 5.6 |
| 5 to 9 years | 20,011 | 6.0 |
| 10 to 14 years | 21,821 | 6.6 |
| 15 to 19 years | 21,824 | 6.6 |
| 20 to 24 years | 21,383 | 6.4 |
| 25 to 29 years | 22,100 | 6.7 |
| 30 to 34 years | 22,979 | 6.7 |
| 35 to 39 years | 22,371 | 6.4 |
| 40 to 44 years | 21,362 | 6.0 |
| 45 to 49 years | 19,782 | 6.3 |
| 50 to 54 years | 20,891 | 6.4 |
| 55 to 59 years | 21,141 | 6.5 |
| 60 to 64 years | 20,674 | 6.5 |
| 65 to 69 years | 18,352 | 5.5 |
| 70 to 74 years | 15,426 | 4.6 |
| 75 to 79 years | 9873 | 3.0 |
| 80 to 84 years | 6279 | 1.9 |
| 85 years and over | 5963 | 1.8 |
| Total | 331,894 | |

*Source* US Census Bureau/American Community Survey

Age-based cohorts may also be carved out in order to focus on specific sub-populations such as teenagers (i.e., 13–19), child-bearing age women (i.e., 15–44), and the oldest-old (i.e., 85 and older). It is important to examine the age distribution of a population because of the importance that various ages have for the nature of that society. For example, in the 1950s "teenagers" emerged as a distinct age cohort in US society. Today, our society is heavily influenced by the large (and growing) number of senior citizens.

Means and medians are often used as summary indicators of the overall age distribution. The mean age (or arithmetic average) of the population is sometimes used as an indicator of age distribution, although the mean is much more sensitive to extreme values than the median and, thus, considered less meaningful.

The median age represents the year at which half of the population is above and half is below. This metric is most commonly used since it provides the best indicator of the mid-point of the age distribution. The median age is usually considered in the light of the age distribution, since the median could reflect a variety of combinations reflecting the relative importance of various age cohorts. Box 4.2 depicts the trend in the median age for the US population from 1960–2020.

Demographers also calculate the mode when examining the age distribution for a population. The modal age cohort indicates where the largest number of people are concentrated.

**Box 4.2: Median Age: United States 1960–2020**

| *Year* | *Median age* |
|---|---|
| 2020 | 38.6 |
| 2010 | 37.2 |
| 2000 | 35.3 |
| 1990 | 32.4 |
| 1980 | 30.0 |
| 1970 | 28.1 |
| 1960 | 29.5 |

*Source* US Census Bureau

## *Sex Ratio*

The *sex* or *gender* of an individual is perhaps the most straightforward attribute to determine since historically there were thought to be only two possible options—male and female. However, it has become increasingly clear that the situation is more complicated than traditionally believed. For our purposes, we will focus on those born male and female (cisgender).

The sex distribution is typically presented in terms of raw numbers (e.g., 5200 females and 4800 males), percentages (e.g., 52% female and 48% male), or converted into a "sex ratio". The sex ratio indicates the number of males per 100 females. Based on the raw numbers above, a sex ratio of 92.3 would be generated, meaning that for this population there are only around 92 males for every 100 females. In most developed countries the sex ratio is less than 100, indicating fewer males than females.

The overall sex ratio figure may mask significant differences within a population. In fact, different sex ratios are exhibited for different age cohorts, with younger age groups in the US exhibiting a higher sex ratio (since more males are born than females) and older age groups exhibiting a lower sex ratio (since females outlive males on the average. Thus the sex ratio at birth in the US today is 103 (i.e., 103 males born for every 100 females). As the population ages, the sex ratio decreases due to the higher attrition rate for males. For the age range 85 and over there are two women for every man (or a sex ratio of 50). The sex ratio is likely to vary for different racial and ethnic groups in society as well. Today, the sex ratio for the non-Hispanic white population is much lower than that for various non-white groups, explaining the fact that more than half of the nation's school children are minority group members.

Race and Ethnicity. Race and ethnicity are at the same time biologically determined and socially constructed. Racial identity is based on physical characteristics such as skin color. Ethnic identification, on the other hand, is based on a common cultural heritage. Both "race" and "ethnicity" are social constructs with race, in particular, having no scientific basis. Thus, the number of racial groups and the basis for racial categorization varies from society to society. In the US racial groups include whites, African-Americans, Asian-Americans and American Indians (including Alaska natives). The decennial census is the major source of data on the racial composition of the US population and in recent censuses it has been possible to claim two or more races.

***Race and ethnicity are at the same time biologically determined and socially constructed.***

Ethnic identification may be determined by members of a subgroup or ascribed by the larger society. Members of many ethnic groups view themselves as distinct from the larger society; others may not see themselves as different at all. Thus, while the Census Bureau may categorize all residents of Spanish origin as "Hispanic," few Mexicans, Cubans or Puerto Ricans would apply that label to themselves.

The only ethnic group officially recognized by the Census Bureau is Hispanics, although there are other ethnic groups within US society that could be identified. Jews and Arabs could both be considered ethnic groups in US society as could subgroups identified based on their national heritage (e.g., Southeast Asians, east Indians, Ukrainians). In fact, the Census Bureau is in the process of establishing a new racial/ethnic category—Middle Eastern/North Africa (MENA). It remains to be seen whether this will be considered a racial group or an ethnic group. Of course, to be considered as an ethnic group, the subpopulation must retain significant aspects of its traditional culture.

When the US population's racial/ethnic composition is examined, Hispanics are classified as ethnic group members as well as in terms of their race. Box 4.3 presents a breakdown of the current racial and ethnic distribution for the US population. Note that the total exceeds 100% since Hispanic-origin persons are double counted—i.e., as members of an ethnic group also assigned to various racial groups. The data show that more than one-third, 35.6%, of the US population is made up of persons classified as a racial or ethnic minority. In fact, these proportions have been rising and will continue to increase given the differences in age structure and fertility behaviors across all racial and ethnic groups.

**Box 4.3: Racial and Ethnic Composition of the US Population: 2020 (in Thousands)**

| *Race/ethnicity category* | *Number* | *Percent* |
|---|---|---|
| Total population | 304,060 | |
| White | 223,553 | 70.4 |
| Black or African American | 38,929 | 12.6 |
| American Indian, Alaska native | 2932 | 0.8 |
| Asian alone | 4674 | 5.6 |
| Native Hawaiian and Pacific Islander | 543 | 0.2 |
| Two or more races | 9009 | 5.2 |
| Some other race | | 5.1 |
| *Hispanic origin* | *50,478* | *18.2* |

*Source* US Census Bureau (American Community Survey)
"Hispanic origin" is italicized since Hispanics are double-counted--once for race (white) and again for ethnicity (Hispanic). The italics is meant to explain why the total adds up to more than 100%.

## Displaying and Analyzing Compositional Variables

Data on compositional variables will be typically generated as raw data, and it is up to the analyst to convert these data into meaningful information. Since different geographies will record populations of differing sizes, it is difficult to compare one geography or population to another using raw data. Therefore, it is necessary to convert raw data into a form that allows for meaningful comparison. For example, knowing that 10,000 deaths occurred in Florida and 2000 deaths occurred in North Dakota in 2005 does not really allow us to compare the health status of the respective states. But, if we convert these into death rates and generate a crude death rate of 8.5 per 1000 population for Florida and a rate of 9.5 per 1000 for North Dakota, we have a basis for comparison.

***In comparing the raw data for one geography or population with another, it is usually necessary to convert the raw data into rates.***

Compositional traits are often expressed in proportion terms (e.g., 18% of the population in Shelby County, Tennessee, is of Hispanic origin) along with summary measures such as means and medians. The use of any of these summary measures can result in useful insights. When using percentage distribution information, it is generally better to include data for several categories (perhaps the entire distribution), even though the focus may be on only one level of aggregation. This will give the user/reader a more complete picture of the attributes being addressed.

Since the mean refers to the arithmetic average and the median to the midpoint of a distribution, these two measures of central tendency are likely to be used for different purposes. It is generally better to rely on the latter, though the use of both of these statistics, along with an examination of the entire distribution, is the best approach. A given mean or median may be the result of an infinite number of combinations of distributional data. Therefore, utilizing the mean or median without analyzing the distribution from which those figures are derived may result in an incomplete understanding of the data.

## Population Pyramids

Population distributions are sometimes presented visually in the form of *population pyramids*. The age/sex distribution of a population is presented in a series of stacked bars. Each bar represents the number or the percentage of the population at that age cohort. The left side displays the percentage (or number) of males in each age cohort and the right side the percentage of females in each age cohort. Box 4.4 presents the population pyramid for the United States in 2020.

**Box 4.4: Population Pyramid United States 2020**

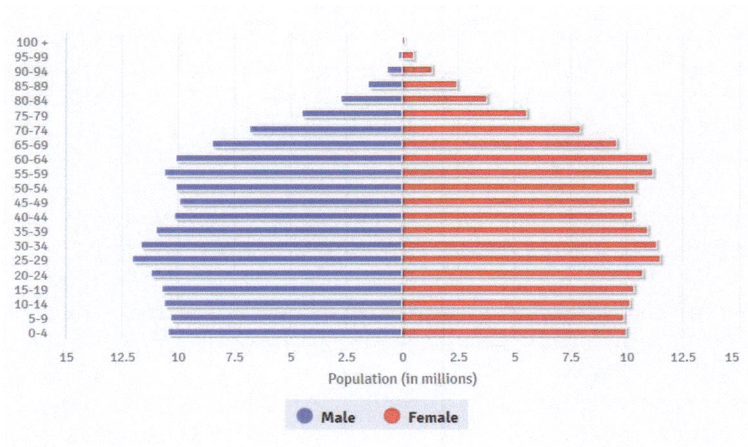


*Source* US Census Bureau

The pyramid provides useful insights into the current age-sex distribution for the US population. Examining this pyramid, we find that there are more males at the younger ages than females; conversely, there are more females at the older ages than males. Over time the disparity at the top can be expected to increase as the US population continues to age. Pyramids that reflect younger age structures are "bottom heavy" have younger age structures (as in Box 4.5 below), while older age structures demonstrate more constant age-to-age percentages.

***Population pyramids indicate a lot about a population and are useful for comparing one population to another.***

The pyramid is also useful for tracking population trends over time. The bulge in the pyramid a little more than halfway up reflects the large baby boom cohort born between 1946 and 1967. The indention below reflects the baby bust when there was a precipitous decline in births. The bulge about one-third of the way up represents the huge Millennial cohort born during the 1990s. By tracking the movement of these cohorts through the pyramid over time, it is possible to monitor the "pig in the python" as cohorts pass through various stages.

This is where demographers have an advantage over most other scientists. The population pyramid allows analysts to predict the characteristics of the population for the future. Examining the pyramid, it is possible to determine how many children will be entering the first grade in five years, how many people will be reaching retirement age in 20 years, or how many people will require nursing home in 30 years.

A series of population pyramids could give a clear picture of how a country transitions from high to low fertility rates. If the pyramid has a broad base, this indicates that a relatively high proportion of the population lies in the youngest age band, such as ages 0–14, and that the fertility rate recorded for the country is high and above replacement level. If a population is below replacement level, the older population becomes increasingly dominant due to a decrease in births over time.

**Box 4.5: Population Pyramid Egypt 2020**

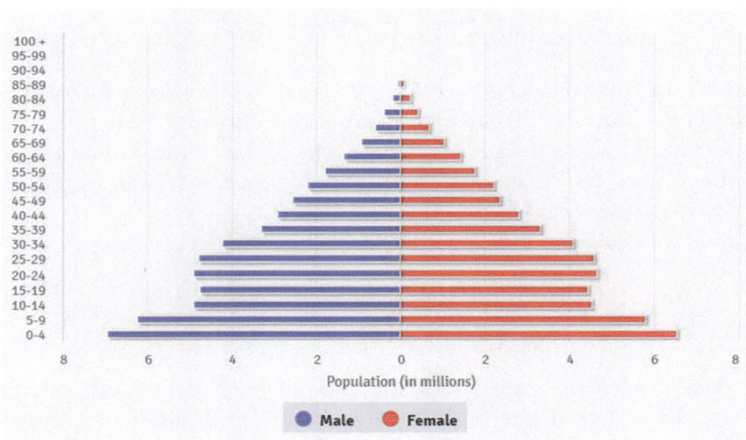


*Source* United Nations

## Dependency Ratios

Age data can be used to calculate *dependency ratios*, or the quotient of an area's dependent population divided by its "supporting" population. Dependent and supporting populations are defined in terms of economic dependence and support. The supporting population in the United States is usually considered those individuals between the ages of 18 (or 20) and 64, while dependent populations are under age 18 (or 20) and over age 64. Dependency and support are general notions regarding economic activity, and the population aged 18 (or 20) to 64 is considered to be economically active (income earning).

The *youth dependency ratio* for the United States in 2020 can be calculated as follows:

$$\frac{\text{number of persons under age 18}}{\text{number of persons 18-64}} = \frac{73,150,000}{197,496,000} = 0.38$$

This ratio of 0.38 converts to 2.70 persons of working age for every person under age 18. Since 1950 this ratio has declined from 0.58 while those of working age have increased proportionately with regard to persons under age 18). This reflects a decline in the proportion of children within the US population.

The *age dependency ratio* for 2020 can be calculated as follows:

$$\frac{\text{number of persons aged 65 and over}}{\text{number of persons 18-64}} = \frac{54,542,000}{197,496,000} = 0.28$$

This ratio of 0.28 converts to 3.60 persons of working age for each person age 65 and over. Since 1950 this ratio has increased from 0.14 (or 7.14 persons of approximately working age for each person 65 and older). The age dependency ratio is used to illustrate the ability of an area's working-age population to support dependent children and senior citizens.

The *total dependency ratio* for 2020 can be calculated as follows:

$$\frac{\text{number of persons aged 65 and over}}{\text{number of persons } 18-64} = \frac{127,793,000}{197,496,000} = 0.65$$

The *total dependency ratio* takes the sum of both dependent populations (under age 18 and over age 65) and divides by the number of persons aged 18 to 64. In 2020, the total dependency ratio was 0.65, or 1.55 persons of working age for every person under age 18 or over age 64. This compares to over 3 workers per dependent in 1950 and reflects the dynamic growth in the elderly population.

***The age dependency ratio is used to illustrate the ability of an area's working-age population to support dependent children and senior citizens.***

## Cohort Analysis

Another way of examining compositional data is through *cohort analysis*. A cohort is a group of persons within a population with a common characteristic or characteristics. Age is the most frequent basis for cohort identification used by demographers. Age cohorts are identified by grouping together persons of similar age. It is assumed, although not always correctly, that persons within a cohort share experiences and behavior because of their common characteristic. Cohort analysis typically involves following a cohort over time to measure the effects of exposure to various events, conditions, etc., such as American soldiers exposed to Agent Orange during the Vietnam War or patients undergoing a particular medical procedure.

***A cohort is a group of persons within a population with a common characteristic or characteristics.***

Cohort analysis is sometimes used as a basis for ascribing experiences and behavior patterns to a cohort when individual data are not available. Cohort analysis can range from measuring the change in number and proportion of persons in various age groupings over time to ascertaining how cohort experiences affect different types of behavior (Swanson and Siegal, 2004). Cohort-to-cohort comparisons allow an assessment of change in cohort behavior over time.

Cohort analysis can be particularly useful in a healthcare setting. Since health problems are frequently age specific, cohort analysis can determine future patterns of morbidity and mortality. Age cohorts also exhibit varying types of health-related behavior or levels of exposure to a particular type of carcinogen that is unique to the population in question. The cohort in question could be tracked over time to provide insights, for example, into disease prevalence. Cohort analysis, in fact, provides the basis for any actuarial exercises such as setting life insurance rates.

Studying the transition or aging of a cohort from one period to the next may involve the use of mortality data and survival analysis. For example, the cohort aged 65 to 69 in 2010 is made up of those persons aged 60 to 64 in 2005 minus those who died during the interval (not accounting for immigration). Tracking change in cohort size tells the analyst a great deal about health service demands that are likely to characterize this cohort in the future. Box 4.6 depicts the structure of a cohort study.

**Box 4.6: Illustration of Cohort Study**

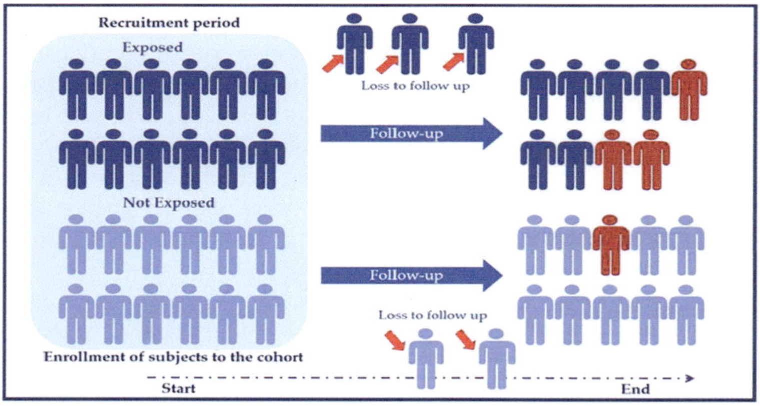


## Data Standardization

Population sizes vary between different areas and, without some way of standardizing the phenomena being evaluated, comparisons between two or more populations can be misleading. Standardized rates can be used to control for differences in population size, allowing for the comparison of data for two or more regions, states, metropolitan areas, or market areas with respect to mortality, morbidity, available resources, and various behaviors. However, even a simple comparison of rates can lead to incorrect conclusions.

***Standardized rates can be used to control for differences in population size, allowing for the comparison of between communities of different sizes.***

For example, if two communities of 10,000 people were being compared and analysts found that one had twice the breast cancer incidence rate as the other, it would be logical to assume that morbidity levels are higher for one population than the other. While variations in morbidity levels may be able to provide some explanation for the observed differential, demographic explanations must be considered as well. The simplest demographic analysis would determine if there were marked sex differences between the two communities. In other words, a heavily female community would clearly be expected to have more cases of breast cancer. Or suppose that one community has a much older age structure than the other. Since breast cancer is more common among older women, the older community would be expected to report more cases of breast cancer.

Up until this point, it has been argued that the age/sex structure differences in the two communities must be studied before cancer incidence is evaluated. Are there any other demographic variables that might help explain why there is a breast cancer rate difference between the two communities? The researcher may want to consider racial-ethnic compositional differences and/or any other demographic factors known to be associated with breast cancer (e.g., child-bearing history). The ultimate goal is to hold constant or control for as many of these factors as possible in order to eliminate competing explanations.

Box 4.7 illustrates the procedure for adjusting for age and sex differences. Both communities are equal in size and have the same incidence of breast cancer. However, as seen in the table, Community A has 50,000 females while Community B has 60,000 females. Because breast cancer is very rare in males, it makes better sense simply to calculate a female-specific rate.

However, suppose that Community A had the same sex structure as that for B. How many breast cancer cases would there be? The rate, 10 per 1000, would have to be multiplied by the size of the female population in Community B, or 60,000. The result of 600 cases is 20% greater than the 500 cases observed. The same adjustment can be made in the other direction; that is, the number of expected breast cancer cases in Community A (assuming it has the same rate as that of Community B) can be calculated. Multiplying 8.33 per 1000 by 50,000 yields 416 cases, a 17% reduction from the 500 cases observed. The new rates per 1000 females for communities A and B are 10.0 and 8.33, respectively. If Community A had the same age structure as Community B, there would have been 638 breast cancer cases, an increase of about 6.3%. The new "adjusted"rate is 638/60,000, or 10.63 per 1000.

**Box 4.7: Age/Sex Composition and Incidence of Breast Cancer for Two Fictitious Communities**

| *Age group* | *Community A* | | | | *Community B* | | | |
|---|---|---|---|---|---|---|---|---|
| | *Males* | *Females* | *Cases*[a] | *Rate*[b] | *Males* | *Females* | *Cases*[a] | *Rate*[b] |
| 60–64 | 13,000 | 13,000 | 100 | 7.69 | 10,000 | 12,000 | 100 | 8.33 |
| 65–69 | 12,000 | 12,000 | 75 | 6.25 | 10,000 | 12,000 | 100 | 8.33 |
| 70–74 | 10,000 | 10,000 | 75 | 7.50 | 12,000 | 14,000 | 100 | 7.14 |
| 75–79 | 8000 | 8000 | 125 | 15.42 | 10,000 | 12,000 | 100 | 8.33 |
| 80 and over | 7000 | 7000 | 125 | 17.86 | 8000 | 10,000 | 100 | 10.00 |
| Total | 50,000 | 50,000 | 500 | 10.0 | 40,000 | 60,000 | 500 | 8.33 |

[a]Cases of breast cancer
[b]Rate per 1000 female population

From an applied standpoint, the analyst may have hypothesized that health conditions were the same in each community, given that each had the same "rate" of breast cancer. However, after adjusting for age and sex, one could conclude that health conditions in Community A are more favorable than in Community B. Of course, other factors would have to be considered before the analysis was complete.

## Sources of Data for Biosocial Variables

There are three main sources of data on biosocial variables: government agencies, professional associations, and commercial data vendors. Governments at all levels are involved in the generation, compilation, manipulation and/or dissemination of demographic data. The federal government, through the decennial census and related activities, is the world's largest processor of demographic data, and the Census Bureau is the primary source of most demographic data for the US. Other federal sources such as the Bureau of Labor Statistics (e.g., occupations), the Department of Health and Human Services (e.g., insurance status), and the Department of Education (e.g., educational achievement) contribute to the data available on the composition of the US population).

State and local governments are also major sources of demographic data. State governments generate a certain amount of demographic data, and each state has a population center for generating and disseminating demographic data. Data centers at universities may also be involved in the generation and processing of demographic data. City or county governments may produce local population projections and other demographic data.

Many private organizations repackage data collected elsewhere (e.g., from the Census Bureau or the National Center for Health Statistics) and present this information within a specialized context. The Population Reference Bureau, a private not-for-profit organization, distributes population statistics in various forms, for example. Some organizations, like the American Association of Retired Persons (AARP), not only compile and disseminate secondary data but are actively involved in primary data collection, as well as the sponsorship of numerous studies.

A recently available resource of data on sociocultural attributes is policymap.com. This not-for-profit organization has pulled together data for the entire United States from a variety of sources and integrated them into a database that allows for data extraction and mapping. The database combines data from various government organizations (although heavy on Census Bureau data) and presents this information along with data derived from non-governmental sources. Data are presented for most levels of geography, often down to the census block group level. There is no charge for accessing any of data sets and a minimal charge for full access to the database's functionality.

A similar compilation of data is offered by city-data.com. Its database pulls together data from a variety of sources and presents them in easily understood form. City-Data is geared toward a more commercial audience and includes data on business activity, real estate development and other non-demographic data.

Commercial data vendors have emerged to fill perceived gaps in the availability of various categories of demographic and health data. These include commercial data vendors that establish and maintain their own proprietary databases, as well as those that reprocess and/or repackage existing data. Because of the demand for demographic data, several commercial data vendors have added demographic data to their inventories. These vendors not only repackage existing data into more palatable form, but some have developed their own proprietary databases. Some data vendors conduct major nationwide surveys of consumers.

## Key Points

- Compositional variables are useful for profiling a population and give a population its particular "personality"
- Compositional variables include biosocial attributes (the subject of this chapter) and sociocultural attributes (the subject of Chap. 5)
- For many purposes, the age structure of a population represents the most significant compositional variable.
- Race and ethnicity are at the same time biologically determined and socially constructed
- In comparing the raw data for one geography or population with another, it is usually necessary to convert the raw data into rates
- Population pyramids indicate a lot about a population and are useful for comparing one population to another.
- The age dependency ratio is used to illustrate the ability of an area's working-age population to support dependent children and senior citizens
- A cohort is a group of persons within a population with a common characteristic or characteristics, and the cohort study is a useful tool for demographers
- Standardized rates can be used to control for differences in population size, allowing for the comparison of between communities of different size

## Discussion Questions

4.1. How is the US population pyramid expected to change over the next 20 years? What will its configuration be? What about the ratio of men to women? What has been the experience of other countries who are further along in the aging process?

4.2. What are the implications for American society of the shifting racial and ethnic distribution? As the non-Hispanic white population shrinks in number and minority population expand in number what changes can we anticipate for US society? Do you consider this growing diversity a positive development of a negative development?

## Exercises

*Exercise 4.1: Population Pyramid*

For this exercise, students should review the population pyramids displayed in Boxes 4.4 and 4.5. The two versions of pyramids should be compared and any similarities and differences noted. At a minimum, students should describe what we can tell about births and death rates and what we might surmise about the median ages of these two populations.

*Exercise 4.2: Calculating Dependency Ratios*

For the three populations below students should calculate the total age-dependency ratios. The ratio for each community should be specified along with the number of workers available per dependent member to support the dependent segment of the population.

| | *Population by age cohort* | | |
|---|---|---|---|
| | *0–17* | *18–64* | *65 +* |
| County A | 2000 | 7000 | 1000 |
| County B | 1000 | 7000 | 2000 |
| County C | 2000 | 6000 | 2000 |

## Reference

Swanson, D. A., & Siegel, J. S. (2004). *The methods and materials of demography* (2nd ed.). Academic Press.

## Additional Resources

Annie E. Casey Foundation. (2019). *US child population grows and changes: 74 million kids to shape future of more diverse America*. aefc.org.

European Environment Agency, Population pyramids for Europe, Africa and Asia. https://www.eea.europa.eu/data-and-maps/figures/population-pyramids-for-europe-africa

US Census Bureau. census.gov; data.census.gov

CHAPTER 5

# Population Composition: Sociocultural Characteristics

**Abstract** This chapter focuses on the dimension of population composition that includes sociocultural characteristics. The various sociocultural attributes that are used to describe populations include marital status and household structure, educational characteristics, economic conditions (including poverty), occupation and industry, along with other social attributes. Trends in the sociocultural attributes of the US population are reviewed in this chapter and the implications of the changing nature of its sociocultural profile described.

## Introduction

*Sociocultural factors* reflect the position of society members within the social structure. Sociocultural factors, in US society at least, are primarily "achieved" rather than ascribed. These are not traits one is born with in a biological sense but those that are acquired (voluntarily or involuntarily) through one's place in the social system. These factors are "cultural" in that those affected take on characteristics assigned by society. Sociocultural factors include marital status, income, education, occupation, and religion among others. Each variable is discussed in turn in the sections that follow.

Marital Status/Living Arrangements/Family Structure. Marital status, living arrangements, and family structure are all ways of looking at household characteristics. One or more of these has served to define the nature of the American population at one time or another. As will be seen, a great deal of diversity now characterizes the US population in regard to these attributes. In the past, marital status was thought by demographers to be the best indicator of household relationships. However, as what we think of as the "traditional"



R. K. Thomas, *Demography: An Introduction to Population Studies*, Springer Texts in Social Sciences, https://doi.org/10.1007/978-3-031-56623-3_5

family gave way to new and different types of households, other measures of household characteristics became more salient.

***Marital status, living arrangements, and family structure are all ways of looking at household characteristics.***

A *household* is defined as one or more persons living in a housing unit. A *housing unit* is defined as one or more rooms that comprise separate living quarters with access from the outside or through a common hall as well as a kitchen or cooking equipment for exclusive use. Thus, a household represents the people who live in a housing unit. The terms "household" and "housing unit" are often used interchangeably. A *family* is defined as two or more persons related by blood, marriage, or adoption who live together. Two persons living in the same housing unit are considered to be a household regardless of the relationship; if they are related to each other they are classified as both a household and a family. Demographers make a distinction between family households and non-family households. *Group quarters* refers to any congregate living arrangement and includes dormitories, nursing homes, prisons, military barracks, assisted living facilities, and residential treatment centers.

## *Marital Status*

Individuals are typically grouped into four marital status categories: single, married, widowed, and divorced. The Census Bureau also recognizes a "married but separated category," although this does not constitute a reportable marital status in all states. In addition to the current marital status of individuals, information may be collected on previous marriages and selected characteristics of any marriage (e.g., age at marriage, race/ethnicity of bride and groom, etc.)

Historically, most Americans married when they entered adulthood and virtually everyone eventually married. However, beginning in the 1960s the rate of marriage began to drop, and this decline has continued to the point that barely one-half of American adults are married today. While the majority of adults are married, this figure has declined sharply over time, while the proportions never married, widowed and divorced have increased. Box 5.1 shows the current marital status of the US population. This is discussed in more detail later in the chapter.

An additional wrinkle has been added to the study of marital status with the emergence of official same-sex unions. Once legally forbidden in the US (and most other countries), as of 2015 same-sex marriages (by Supreme Court ruling) became legal in all 50 states. Although same-sex marriages still account for a small number of total marriages, their numbers have increased since this marital status category was first recognized in the 1980s. In 2020 it was found that there are over one million same sex married couples in the US (accounting for less than 1% of all married couples). There are more female married couples than males, and married same-sex couples now represent a majority of all same-sex couples.

**Box 5.1: Marital Status of the US Population 15 Years and Over: 2020**

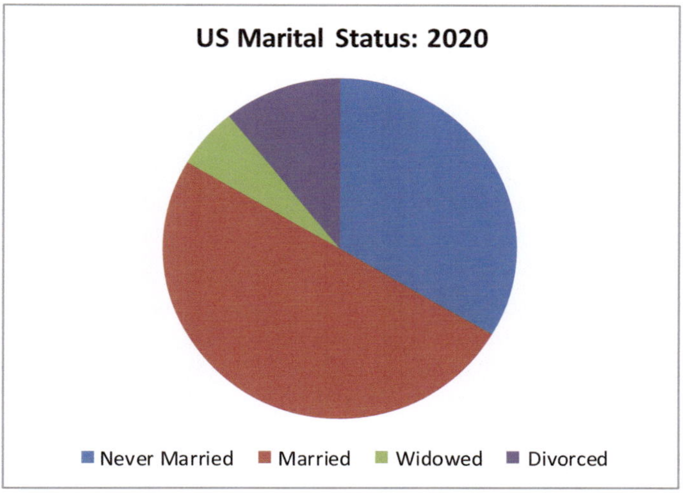


*Source* US Census Bureau

The distinctions between different types of household structures are important for a number of reasons. Family households, for example, have legal standing, while nonfamily households typically do not, although recent changes in state laws in this regard should be noted. Further, family households are likely to differ from nonfamily households in a number of ways unrelated to the size or nature of the relationship. (Box 5.2 presents the standard Census Bureau breakdown of household types.)

***A family is defined as two or more persons related by blood, marriage, or adoption who live together.***

The proportion of households considered family households (two or more persons who are related) has declined while the number of female- and male-headed households without a spouse present (single-parent households) have increased proportionately. Moreover, there has been a rise in the proportion of single-person households over the past quarter century.

**Box 5.2: Household Composition of the United States: 2020 (in Thousands)**

| | | |
|---|---|---|
| Number of Households | 122,354 | |
| Persons per household | 2.60 | |
| Persons per family household | 3.21 | |
| Family households | 83,677 | 68.3% |
| Married couple | 58,807 | 48.1% |
| Male-headed households (no spouse) | 21,780 | 17.8% |
| Female-headed households (no spouse) | 33.608 | 27.5% |
| Nonfamily households | 44,744 | 38.7% |
| One-person households | 36,200 | 28.0% |

*Source*: US Census Bureau

Increasingly, Americans are residing in congregate housing. While facilities like prisons, college dormitories and residential treatment centers are fairly stable in numbers, there has been a steady increase in senior living facilities and assisted care facilities. These numbers are expected to continue to grow as the US population ages. There are a few miscellaneous categories of dwellings such as house trailers, motor homes and boats. Of these, only house so-called "mobile homes" exist in any number.

### *Group Quarters*

One additional residential category to be considered is group quarters. Group quarters are defined by the Census Bureau as living arrangements for groups not living in conventional housing units or groups living in housing units containing ten or more unrelated people or nine or more people unrelated to the person in charge. Sometimes referred to as "congregate housing," group quarters are owned or managed by an entity or organization providing housing and/or services for the residents. These services may include custodial or medical care as well as other types of assistance. Residency is commonly restricted to those receiving these services. People living in group quarters are usually not related to each other. Group quarters include such places as college residence halls, residential treatment centers, skilled nursing facilities, assisted living facilities, group homes, military barracks, correctional facilities, and workers' dormitories.

***Group quarters (or "congregate housing") are owned or managed by an entity or organization providing housing and/or services for the residents.***

### *Housing Classification*

Americans reside in a variety of housing types. These range from one-room rentals to congregate facilities with hundreds or thousands of residents. In the popular mind, the typical type of housing is a single-family detached dwelling (typical of suburban communities and rural areas). This type of dwelling is representative of the "American dream" but not nearly as common as generally thought. Another category involves single-family *attached* dwellings that include a variety housing types. These include duplexes, quadraplexes, condominiums and apartment complexes of various sizes. These are distinguished by the number of units in a complex, with apartments often classified from a few units up to hundreds of units. Condominiums, of course, take the form of apartments, with the occupants actually owning their units rather than renting them.

As will be demonstrated later, demographers (and others) who generate population projections often make use of housing data as a component of their analyses.

Measures of Economic Status. Because of the importance of economic standing in American society, a significant amount of attention is paid to income and other measures of wealth. *Income* refers to the amount of money taken in by individuals and households during a specified time period (usually a year). Income statistics generally refer to income in the previous year with income reported in absolute dollars. In most cases, income data are collected for the household, whether it is a family household or a group of unrelated individuals (i.e., family income vs. household income).

One other measure of economic status is labeled as *wealth*. This refers to the amount of resources that one has at their disposal (i.e., net worth). Unlike income, wealth is not measured in terms of revenue generated during a year's time but is calculated based on the accumulated wealth of an individual. While there is significant variation in income for various segments of society, the "wealth gap" is considerably greater.

In addition to the amount of income, data may be collected on the source of income. Most Americans generate income by working for a salary or an hourly rate. Large numbers also generate income by entrepreneurial activities like business ownership. Others generate revenue through investments or through interest earned on various securities. A growing number of Americans receive government payments. While the majority of these are Social Security recipients, additional numbers of citizens receive welfare payments or disability compensation from a government agency.

### *Presenting Income Data*

Analyzing the income distribution for over 330 million Americans is a daunting task for demographers, so income data are using presented in some aggregate form. Incomes are typically group into categories rather than being viewed as single cases. Thus, the distribution of income across a continuum is a typical way of viewing its distribution by demographers. A simple approach might involve grouping incomes into ten-year intervals—e.g., $0 to $9,999, $10,000 to $19,999, $20,000 to $29,999 and so forth. Even that grouping is a little cumbersome, so it is usual to employ broader categories—e.g., $0 to $14,999, $15,000 to $24,999, $25,000 to $49,999 and so forth. As US incomes have increased over the decades, the income categories have generally become larger.

It is also common to generate the same calculation of only family households, yielding a median household income. These represent aggregated income statistics. The median is determined based on the income level at which half of the population is below that figure and half is above (or the mid-point). Box 5.3 present summary income statistics for the US.

**Box 5.3: Summary Income Statistics United States 2020**

| *Indicator* | *Statistic* |
|---|---|
| Median household income | $64,994 |
| Mean household income | $91,547 |
| Modal household income | $50,000–$74,999 |
| Median family income | $80,069 |
| Mean family income | $107,335 |
| Modal family income | $100,000–$149,999 |
| Per capita income | $35,384 |
| *Poverty rate* | |
| All people | 12.8% |
| Families | 9.1% |
| Children under 18 | 17.5% |

*Source* US Census Bureau

The income distribution may also be presented in terms of mean income rather than median income. Mean income represents a true average in that it takes aggregate income for all households and divides that by the number of households. Although this may appear to be a statistical sound approach, it suffers from the influence of outliers. Thus, if a population has a large number of households with no income, the mean will understate the general level of income. Similarly, if a population has a small number of households with excessive income, this may skew the mean in other direction.

***The distribution of income for a geographic area is often presented in terms of median household income or median family income.***

One final measure of central tendency is the modal category—i.e., the grouping that has the most households in it. This is primarily useful when the analyst is interested in the distribution of income across the population. It also serves the purpose of clarifying the significance of the median income level.

One other indicator of income level that is frequently used is per capita income. This indicator is calculated differently in that total income for a population is established and then divided by the number of individuals in that population. Per capita income is not considered as useful as household-based indicators because the per capita income can be influenced by a number of factors that might make such an average misleading.

### *Measuring Poverty*

One other indicator of a population's income is its *poverty* level, or the extent to which individuals, families or populations are economically deprived. Poverty can be measured in absolute or relative terms, depending on the intent. In the US the federal government establishes the criteria for what is considered to be the "official" poverty threshold. These thresholds are based on the amount of money required to cover basic living expenses (e.g., housing, food, clothing). The Census Bureau uses a table of income categories that vary by family size and composition for assigning poverty status. If a family's total income is less than the family's threshold, then that family and every individual in it is considered in poverty. The official poverty figures distinguish between urban and rural households in calculating the threshold. The official poverty definition uses money income before taxes and does not include capital gains or noncash benefits (such as public housing, Medicaid, and food stamps) in its calculations.

***It is impossible to address income-related attributes in the united states without considering the existence of poverty.***

In 2020, the poverty threshold for an urban family of four was around $23,000 and an estimated 13% of Americans lived at or below the poverty level at that time. One could think in terms of relative poverty of "statistical" poverty. Statistical poverty refers to efforts (typically on the part of the federal government) to empirically determine what the poverty level is and, subsequently, who falls into that category. The standard approach involves some formula that estimates the basic cost of living for an individual and then sets that as the "poverty guideline." Thus, in 2020, the poverty level established by the federal government for a family of four was $26,200. This figure varies based on family size and on the part of the country being considered.

Relative poverty is measured in terms of the household's relationship to the median household income. Thus, if the nation's median household income is $50,000, a household with an income of $25,000 would be considered to be 50% below the median. Even that approach is somewhat misleading in that it does not reflect the "relative deprivation" that exists among various subpopulations within US society.

***Poverty can be viewed in either relative or absolute terms.***

A great deal of interest is expressed with regard to the poverty level, and many consider it a significant societal problem. Indeed, the facts that in 2020 the federal government calculated that 13% of Americans lived at or below the poverty level. More disturbingly, it was determined that 18% of children were so classified. That means that one out of every 9 people and one out of 6 children live in poverty.

Education. *Education* refers to the attributes of a population as they relate to educational experience. This is an important attribute in American society because of its emphasis on academic achievement and the relationship between educational attainment and income. The educational status of a population is typically stated in terms of the number of school years completed and/or the types of degrees earned. Thus, the American Community Survey solicits information on the highest degree that respondents have earned (e.g., high school diploma, graduate degree). Box 5.4 presents the educational attainment breakdown for the United States in 2020.

While years of education completed describes a population's past academic experience, figures related to current educational enrollment provide insights into a population's current level of educational involvement. This is typically presented as the percent of the population enrolled in preschool, elementary school, middle school, and so forth.

**Box 5.4**

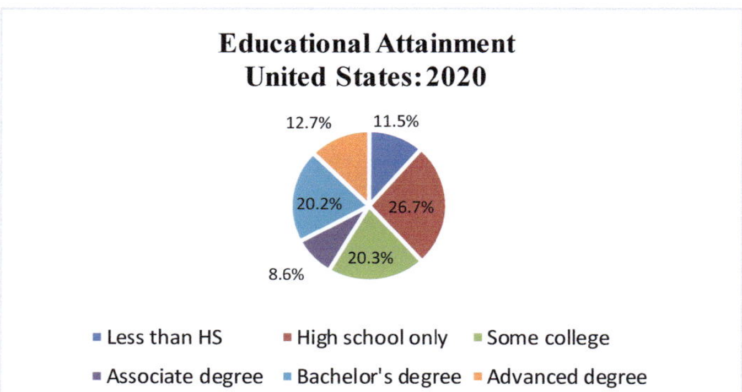


*Source* US Census Bureau

## Work-Related Attributes

*Work status*, *occupation*, and *industry* data all relate to one's position in the labor force. Because of the importance of economic activity in the American ethos, significant effort is expended in characterizing and monitoring work-related activities. This is essentially the only category of demographic attributes that is tracked on an almost real-time basis. Significant government resources are devoted to monitoring and influencing the course of economic activity that this category reflects.

### *Labor-Related Status*

An understanding of work-related activities begins with a discussion of labor force participation. The number and proportion of the population that is included in the labor force has significant implications for the operation of the economic system. Thus, the federal government maintains up-to-date figures on the number of people actively in the labor force. This figure typically includes only those in the civilian labor force. People are considered to be in the labor force if they are currently working (in any capacity) or are looking for work. Related demographic attributes that are also examined that characterize the labor force include the age, sex, race and citizenship status among others of labor members. These figures apply only to those 16 years old or older.

***The number and proportion of the population in the labor force has significant implications for the economic system.***

The size of the labor force is also important in determining the unemployment rate. The unemployment rate is calculated using the number in the labor force as its denominator. Since the labor force includes those who are looking for work, this provides the basis for calculating the unemployment rate. Those who are not in the labor force are not considered for this calculation. Those not in the labor force include housewives, students, retirees, disabled individuals, and those otherwise employable that have removed themselves from the labor force for various reasons.

For individuals who are employed, additional information on the number of hours worked may be collected. Part-time and full-time classifications may be used based upon the number of hours per week and weeks per year worked. While the term underemployment is frequently used today, there is no commonly agreed upon definition for that employment status.

Over time, there has been increasing interest in data on female labor force participation and employment. Women now constitute essentially half of the labor force and controversy over equal pay for equal work remains an issue in

US society. Also at issue is the fact that most women in the work force have young children that need to be cared for. Box 5.5 provides selected labor force statistics.

**Box 5.5: Labor Force Statistics* United States 2020**

| | *Number* | *Percent Eligible population* | *Percent Labor force* |
|---|---|---|---|
| Total eligible population | 216,750 | 100.0 | |
| In labor force | 165,903 | 63.4 | |
| In civilian labor force | 164,759 | 63.0 | |
| Employed | 159,889 | 59.6 | 94.6 |
| Unemployed | 8,871 | 3.4 | 5.4 |
| Not in labor force | 95,747 | 36.6 | |
| Females eligible population | 134,001 | 61.8 | |
| Females in labor force | 78,484 | 56.8 | 47.6 |
| Females employed | 74,173 | 55.8 | 46.4 |

*Population 16 years and older
*Source* US Census Bureau

## *Occupation*

*Occupation* refers to the kind of work a person normally does (that is, the "job"). Examples of specific occupations include registered nurse, gasoline engine assembler, and teacher's aide. This attribute refers to what the worker does rather than where he/she does it. Each occupation is assigned a code from the dictionary of occupational titles (DOC), and individual workers are assigned to an occupational category by the Census Bureau. The large number of occupations is aggregated into a small number of major groupings, such as professional and technical, sales, and management.

The federal government (and some demographers) have a particular interest in the types of occupations extant in the nation's economic sector. The distribution of occupants provides information on the nature and, to a certain extent, the "health" of the economy. As our economy has evolved, the proportional distribution of occupations has continued to change.

The occupational structure includes thousands of different jobs and for reporting purposes the federal government groups occupations into five major categories: management/business/science/arts; service occupations; sales and office occupations; natural resources/construction/maintenance; and production/transportation/material moving. There are hundreds if not thousands of different occupations within these categories. Box 5.6 presents the distribution for major occupational categories for 2020.

**Box 5.6: Employment by Occupational Category* United States 2020 (in Thousands)**

| *Occupational category* | *Number* | *Percent* |
|---|---|---|
| Management/business/sciences/arts | 62,526 | 39.5 |
| Service occupations | 27,095 | 17.4 |
| Sales and office occupations | 33,248 | 21.3 |
| Natural resources/construction/maintenance | 13,620 | 8.7 |
| Production/transportation/material moving | 20,398 | 13.1 |
| Total | | |

*Population 16 years and older
*Source* US Census Bureau

### *Industry*

*Industry* refers to the business or industry where the occupation resides. For the examples above, the registered nurse would be assigned to health and social services, the gasoline engine assembler to manufacturing, and the teacher's aide to educational services. Industries are classified based on the North American Industrial Code (NAIC) system. The distribution of workers by industry provides a profile of the economy, and changes in employment in the respective industries reflect changes in the nature of the economy.

***Occupation refers to the type of work a person does, while industry refers to the sector of the economy where that work is done.***

The federal government divides industries into 13 different categories and allocates workers to these various groups. The industry pertains the sector of the economy in which one works and not the type of work performed. Thus, there is a complicated overlap between occupation designation and industry designation. The occupation of secretary will exist within every occupational category. A computer programmer will be primarily found in the information industry but there is virtually no industrial category in which computer programmers do not work. Aerospace engineers are likely to only be found in the professional/scientific/management occupational category. Box 5.7 presents the current distribution of jobs among different industries.

**Box 5.7: Employment by Industry* United States 2020 (in Thousands)**

| *Industrial category* | *Number* | *Percent* |
|---|---|---|
| Agriculture/forestry/fishing | 2,338 | 1.7 |
| Construction | 10,445 | 6.7 |
| Manufacturing | 15,889 | 10.0 |
| Wholesale trade | 3,897 | 2.5 |
| Retail trade | 17,148 | 11.0 |
| Transportation/warehousing/utilities | 8,573 | 5.5 |
| Information | 3,118 | 2.0 |
| Finance/insurance/real estate | 10,289 | 6.6 |
| Professional/scientific/management | 18,239 | 11.7 |
| Educational services/healthcare/social assistance | 36,322 | 23.3 |
| Arts/entertainment/recreation | 14,654 | 9.4 |
| Other services | 7,483 | 4.8 |
| Public administration | 7,327 | 4.7 |
| Total | 155,889 | |

*Population 16 years and older
*Source* US Census Bureau

Religion. Despite the importance of religion in American society, religion is one of the least studied compositional variables. Questions regarding religious affiliation or one's level of religiosity are not included in censuses and government-sponsored surveys. However, sample surveys, church registries, and even institutional data (e.g., hospital admission forms) may provide partial information on religious affiliation. The questions asked usually concern religious affiliation (e.g., Roman Catholic), attendance (e.g., number of times attending per month), and religiosity (e.g., depth of religious commitment). This topic is of relevance to demographers in that various demographic attributes are associated with different religions.

Religious affiliation is generally presented in terms of different religious faiths and, within this broad category, for specific denominations. This is particularly salient for the US population with its myriad versions of the dominant religion, Christianity. Box 5.8 presents an estimate of the distribution of Americans by religious affiliation.

***While Christian denominations have seen a decline in membership, the numbers of non-Christians and religiously unaffiliated have been growing.***

Historically, Protestant Christians have represented the majority of those affiliated with a specific religion. Catholic Christians continue to constitute the second largest group. Members of non-Christian religious remain small in number compared the historically dominant Christian denominations. The

Jewish religion remains the largest of these although accounting for less than 2% of the total population. Those without religious affiliation have become the third largest group of Americans in terms of religion.

**Box 5.8**

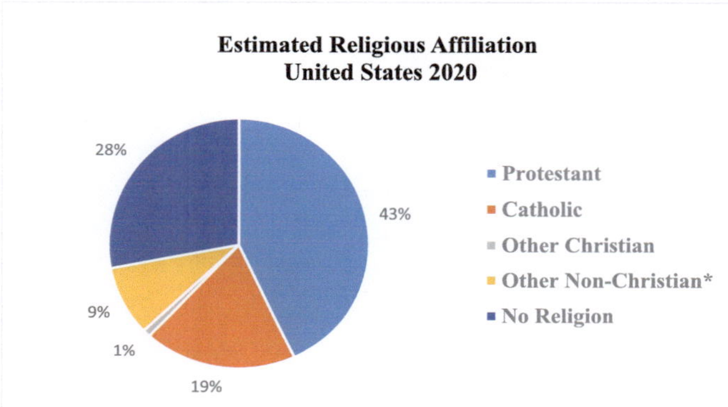


*Other Non-Christian includes: Jews, Muslims, Buddhists, and Hindus

*Source* Pew Research Center

## Trends in Sociocultural Attributes

Key trends related to sociocultural attributes are described below. More information on these trends can be found in Chap. 11.

### *Trends in Marital Status, Household Characteristics and Living Arrangements*

For decades, the family has been undergoing change. First it was high divorce rates, then it was less people marrying (and those who did marrying at a later age); then it was less people having children (and those that did having fewer of them and at a later age). In 2020, the Census Bureau reported that 49% of the US population age 18 and over was married, a low figure compared to the 72% of adults that were married in 1960. In 2020 33% had never married, 11% were divorced, and 6% were widowed. These figures for the non-married all represent record highs and suggest a social milieu much different than that of the twentieth century.

These changes in marital status have had major implications for the US household structure. It has meant that what is popularly considered the "typical" American family (with two parents and x number of children) has become a rarity. According to the American Community Survey, only 18% of the households in 2020 fell into this category. Today, married couple (without children) households have become the most common household form, but this type of household accounts for less than 29% of the total. "Non-traditional" households have become the norm, and an unprecedented proportion of households are one-person households.

**Box 5.9: Trends in Household Types: 1940–2022**

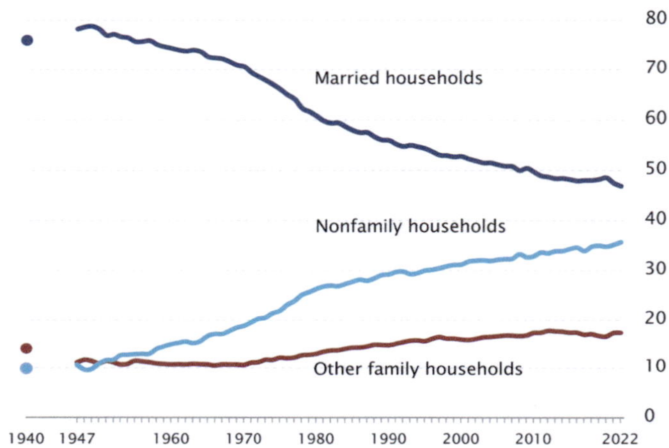


*Source* U.S. Census Bureau, Decennial Census, 1940, and Current Population Survey, Annual Social and Economic Supplements, 1947 to 2022

### *Trends in Economic Status*

Income levels in the US have been steadily rising overall but with some important sub-trends. Median household income in 2020 was over $70,000, a significant increase over the $30,000 recorded in 1990. While this suggests continuing affluence across the board for the American population, this average masks the fact that income (and, particularly, wealth) has become increasingly concentrated within fewer hands. The wealthiest segments of our population have accounted for the bulk of the increase in household income, while the growth for the majority of households has stagnated.

This overall trend masks two distinct episodes in the evolution of household incomes (the first lasting from 1970 to 2000 and the second from 2000 to 2018). Most of the increase in household income was achieved in the period from 1970 to 2000. In these three decades, the median income increased by 41%, to $70,800, at an annual average rate of 1.2%. From 2000 to 2018, the growth in household income slowed to an annual average rate of only 0.3%. If there had been no such slowdown and incomes had continued to increase in this century at the same rate as from 1970 to 2000, the current median US household income would be about $87,000, considerably higher than its actual level of $74,600 (Horowitz et al., 2020). Box 5.10 presents trends in median household income for 1960 to 2018.

**Box 5.10: Median Household Income United States 1985–2020**

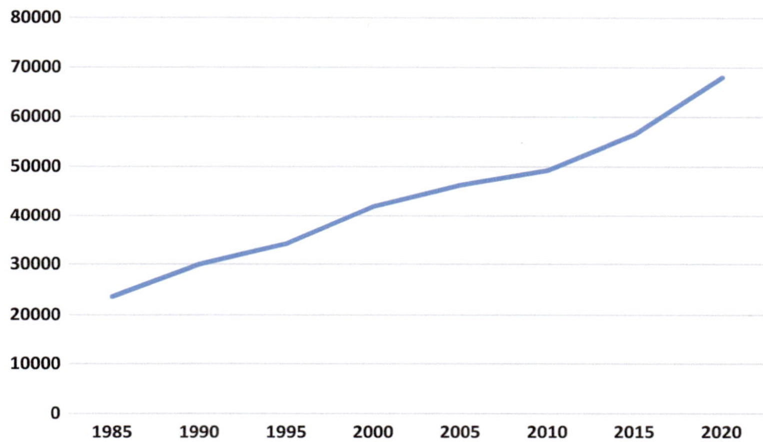


*Source* US Census Bureau

The shortfall in household income is attributable in part to two recessions since 2000. The first recession, lasting from March 2001 to November 2001, was relatively short-lived. Yet household incomes were slow to recover from the 2001 recession and it was not until 2007 that the median income was restored to about its level in 2000 (Horowitz et al., 2020).

Then, household income took another hit due to the COVID-19 pandemic beginning in 2020 and continuing today. The financial hardships caused by the COVID-19 recession in the US were endured mostly by lower- and middle-income families. From 2019 to 2020, the median income of lower-income households decreased by 3.0% and the median income of middle-income households fell by 2.1%. In contrast, the median income of upper-income households in 2020 was about the same as it was in 2019, according to a new Pew Research Center analysis of government data.

The setbacks to the finances of lower- and middle-income households during the pandemic mark a significant reversal from their recent experiences. From 2010 to 2019, following the end of the Great Recession, the median incomes of households in all income tiers had increased at about the same pace—an annual average rate of 1.8% for lower-income families, 1.6% for middle-income families and 1.9% for upper-income families, after adjusting for inflation.

The long-running shift in the distribution of US household income towards upper-income families stayed on track during the coronavirus pandemic. The share of aggregate US household income held by upper-income families reached 50% in 2020, up from 46% in 2010. The share held by middle-income families decreased from 45 to 42% over the same period. The share held by lower-income families also decreased, from 9 to 8% (Kochar and Sechopoulos, 2022).

**Box 5.11: Poverty Rate for Children and Seniors United States 2009–2021**

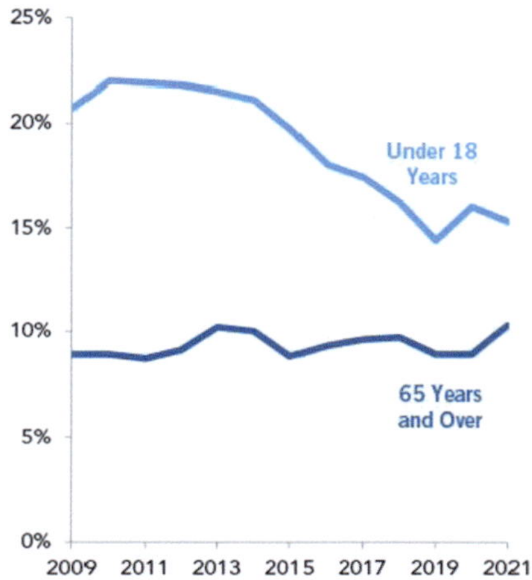


*Source* US Census Bureau

One of the factors affecting income distribution in US society is the changing rate of labor force participation. Historically, there has been a steady increase in the proportion of the nation's adults who were in the labor force. In the 1960s this rate increased dramatically as growing numbers of women joined the workforce. The participation rate leveled off in the 1990s and has been declining ever since. Short-term developments have some impact on labor force participation, but the primary factor today is the aging of the population. The proportion of the population available to work is steadily declining as many workers age out of the workforce. Box 5.12 illustrates labor force participation rates for 1948–2016.

**Box 5.12: Labor Force Participation Rate United States 1950–2021**

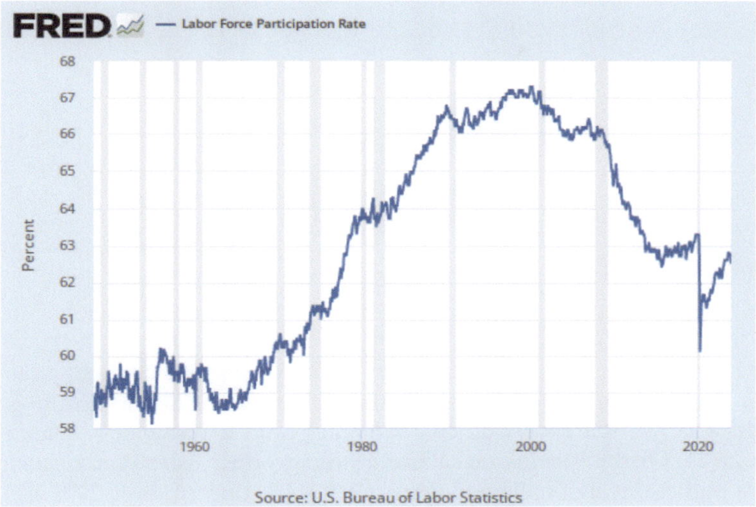


### *Trends in Education*

Over a decades-long period, the US population has become more educated, as increasing numbers of adults completed postsecondary education. In 2021, roughly 26% of the population ages 25 and older had obtained a bachelor's degree. Yet despite this encouraging increase in educational attainment, disparities by race and ethnicity remain high. This is a troubling trend given what we know from research—namely those with a postsecondary education are more likely to be employed, earn higher salaries, own homes, vote, live healthier lifestyles, and volunteer in their communities. These disparities suggest that large groups of Americans are left out of opportunities to improve their lives and to maximize their contributions to society (Espinosa et al., 2020).

The overall educational attainment of adults ages 25 and older increased, as more of the population enrolled in and completed some level of postsecondary education. In 2017, Asians were the racial and ethnic group with the highest levels of postsecondary attainment. Hispanic men and women and American Indian or Alaska Native men had the lowest levels of educational attainment, with more than half having attained a high school credential or less. Across all racial and ethnic groups, adults ages 35 to 44 had the highest levels of educational attainment. However, Hispanics followed a different pattern than any other racial and ethnic group, acial and ethnic group, with each 10-year age cohort having higher rates of college attainment than the next-oldest group (American Council on Education, 2020).

**Box 5.13: Educational Attainment United States 2000 and 2021**

| | *2000 (%)* | *2021 (%)* |
|---|---|---|
| Less than high school | 19 | 9 |
| High school only | 29 | 28 |
| Some college | 21 | 15 |
| Associate degree | 6 | 10 |
| Bachelor's degree | 16 | 24 |
| Advanced degree | 9 | 14 |

*Source* US Census Bureau

Data from 2021 indicate that gender differences in educational attainment exist at each education level when examined for all racial and ethnic groups. Overall, a larger share of women than men hold a bachelor's degree (26% versus 23%). Of the various racial and ethnic groups, Asian Americans exhibited the highest level of college degrees (34%). In contrast, only 17% of African Americans and 15% of Hispanics reported attaining a bachelor's degree.

### *Trends in Religion*

As noted, Protestant Christians have historically represented the majority of those affiliated with a specific religion, with Catholic Christians constituting the second largest group. However, affiliation with these two dominant denominations has been dwindling over time as membership has declined, while small increases in the number of non-Christian and large increases in the number of unaffiliated were recorded.

Although members of non-Christian religious remain small in number compared the historically dominant Christian denominations, the numbers of Buddhists, Muslims and Hindus have been growing in recent years. The Jewish religion has remained at essentially the same proportion over time. Those without religious affiliation have become the third largest group of Americans in terms of religion and the fastest growing of any of the categories. This growth is particularly noteworthy among young Americans. Those remaining in traditional Protestant and Catholic congregations (including African American Protestant denominations) are found to predominantly older citizens.

## Sources of Data for Sociocultural Variables

There are three main sources of data on sociocultural variables: government agencies, professional associations, and commercial data vendors. Governments at all levels are involved in the generation, compilation, manipulation and/or dissemination of demographic data. The federal government, through the decennial census and related activities, is the world's largest processor of demographic data, and the Census Bureau is the primary source of most demographic data for the US. Other federal sources such as the Bureau of Labor Statistics (e.g., occupations), the Department of Health and Human Services (e.g., insurance status), and the Department of Education (e.g., educational achievement) contribute to the data available on the composition of the US population).

State and local governments are also major sources of demographic data. State governments generate a certain amount of demographic data, and each state has a population center for generating and disseminating demographic data. Data centers at universities may also be involved in the generation and processing of demographic data. City or county governments may produce local population projections and other demographic data.

Many private organizations repackage data collected elsewhere (e.g., from the Census Bureau or the National Center for Health Statistics) and present this information within a specialized context. The Population Reference Bureau, a private not-for-profit organization, distributes population statistics in various forms, for example. Some organizations, like the American Association of Retired Persons (AARP), not only compile and disseminate secondary data but are actively involved in primary data collection, as well as the sponsorship of numerous studies.

A recently available resource of data on sociocultural attributes is policymap.com. This not-for-profit organization has pulled together data for the entire United States from a variety of sources and integrated them into a database that allows for data extraction and mapping. The database combines data from various government organizations (although heavy on Census Bureau data) and presents this information along with data derived from non-governmental sources. Data are presented for most levels of geography, often down to the census block group level. There is no charge for accessing any of data sets and a minimal charge for full access to the database's functionality.

A similar compilation of data is offered by city-data.com. Its database pulls together data from a variety of sources and presents them in easily understood form. City-Data is geared toward a more commercial audience and includes data on business activity, real estate development and other non-demographic data.

Commercial data vendors have emerged to fill perceived gaps in the availability of various categories of demographic and health data. These include commercial data vendors that establish and maintain their own proprietary databases, as well as those that reprocess and/or repackage existing data. Because of the demand for demographic data, several commercial data vendors have added demographic data to their inventories. These vendors not only repackage existing data into more palatable form, but some have developed their own proprietary databases. Some data vendors conduct major nationwide surveys of consumers.

## Key Points

- Sociocultural attributes reflect the position of an individual within his social structure.
- Marital status, living arrangements and family structure are ways of looking at household characteristics.
- A family is defined as two or more persons related by blood, marriage, or adoption who live together.
- The changes in family and household characteristics have been so significant that some refer to it as the second demographic transition.
- Group quarters are a special kind of housing arrangement that are managed by an entity providing services to the residents.
- The income distribution for a geographic area is often presented in terms of median household income or median family income.
- The steady increase in median household income disguises the growing gap between the richest and poorest in US society.
- It is impossible to address income-related attributes in the US without considering the existence of poverty.
- Poverty can be viewed in either relative or absolute terms.
- The number and characteristics of the US population has significant implications for the economic system.
- Occupation refers to the type of work done, while industry refers to the sector of the economy where the work is done.
- While Christian denominations have seen a decline in membership, the number of non-Christians and religiously unaffiliated has been growing.

## Discussion Questions

5.1 There have been dramatic shifts with regard to marriage and families in the US. Fewer people are marrying and having children, more babies are born to unmarried women, divorces are increasingly common, and same-sex marriages are on the increase. Students should review the current trends and consider whether this represents, as some argue, a society undergoing social disorganization or a society that is adapting to a changing world. In other words, are these developments a good thing or a bad thing (or neither)?

5.2 Despite the tremendous wealth of our society, large segments of the population remain in poverty. Students should discuss this situation and speculate on why this is the case. What explains persistent poverty in the US? Is this a result of individual behavior or a reflection of structural defects in our society?

## Exercises

*Exercise 5.1: Detailed Demographic Profile*

The project will involve the compilation of relevant data for a county of your choice, along with appropriate description and analysis in narrative form. Data for these attributes should be obtained on-line from the American Community Survey for 2020. The following types of data should be included in table form for the profile:

- Population count
- Population density (along with a discussion of distribution patterns)
- Sex distribution
- Age distribution
- Racial and ethnic characteristics
- Marital status
- Family structure
- Persons per household
- Presence of children
- Income level (including poverty rate)
- Educational level
- Industry characteristics
- Occupational characteristics
- Length of residence

*Exercise 5.2: Detailed Demographic Profile*

The project will involve the compilation of relevant data for a county of your choice, along with appropriate description and analysis in narrative form. Data for these attributes should be obtained on-line from the American Community Survey for 2020. Include any tables, graphs and maps necessary to provide a demographic "snapshot" of the county. With illustrative materials, the assignment should probably be 8–10 pages in length.

The following types of data should be included in the profile:

- Population count
- Population density (along with a discussion of distribution patterns)
- Sex distribution
- Age distribution
- Racial and ethnic characteristics
- Marital status
- Family structure
- Persons per household
- Presence of children
- Income level (including poverty rate)
- Educational level
- Industry characteristics
- Occupational characteristics
- Length of residence

## References

Espinosa, L., Turk, J. M., Taylor, M., & Chessman, H. (2020). *Race and ethnicity in higher education: A status report*. American Council on Education.

Horowitz, J. M., Igielnik, R., & Kochar, R. (2020). *Trends in income and wealth inequality*. https://www.pewresearch.org/social-trends/2020/01/09/trends-in-income-and-wealth-inequality/

Kochar, R., & Sechopoulos, S. (2022). COVID-19 pandemic pinches finances of america's lower- and middle-income families. https://www.pewresearch.org/social-trends/2022/04/20/covid-19-pandemic-pinches-finances-of-americas-lower-and-middle-income-families/#:~:text=The%20financial%20hardships%20caused%20by,income%20households%20fell%20by%202.1%25

## Additional Resources

American Community Survey (data.census.gov).

City-Data Social Network (city-data.com).

Cropper, M., McKibben, J. N., Swanson, D. A., & Tayman, J. (2011). *Vendor accuracy study: 2010 estimates versus census 2010*. ESRI.

Economic Policy Institute (epi.org).

Mayol-Garcia, Y., Gurrentz, B., & Kreider, R. M. (2021). Number, timing, and duration of marriages and divorces: 2016. In *Current population reports*. https://www.census.gov/content/dam/Census/library/publications/2021/demo/p70-167.pdf

PolicyMap (policymap.com).

Population Research Bureau (prb.org).

Smialek, J. (2021). What social trends told us about the US economy in 2022. *New York Times*. https://www.nytimes.com/2021/12/29/business/economy/us-economic-trends-2021.html

CHAPTER 6

# Data Sources for Demography

**Abstract** This chapter describes the types of data that demographers use to describe a population and to address applied problems. The various categories of data are reviewed along with the manner in which these data are generated and the sources from which they can be drawn. This chapter emphasizes the breadth of the data that falls under the aegis of demography and the importance of data for the study of demographic trends.

## Introduction

The grist for the demographic mill is data, and virtually every demographic activity involves the examination of statistics for the population or topic under consideration. The purpose of this chapter is to identify and describe the broad range of secondary data sets of use to demographers as well as others who are interested in population characteristics. While it is not feasible to consider all possible types of demographic data and their sources, this chapter addresses the most important data resources available to demographers, the means through which the data are generated, and the ways in which the data can be accessed.

## Data Collection Methods

The methods of data collection discussed below are divided into four general categories: censuses, registration systems, surveys, and synthetically produced data. Censuses, registries, and surveys are the more traditional sources of demographic data, although synthetically produced statistics such as population estimates and projections have become standard tools for most demographic analyses.



R. K. Thomas, *Demography: An Introduction to Population Studies*, Springer Texts in Social Sciences, https://doi.org/10.1007/978-3-031-56623-3_6

### *Census*

A census involves a complete count of individuals (or entities) residing in a specific place at a specific time. The US Census Bureau (within the Department of Commerce) has conducted population censuses since 1790. Subsequently expanded to the census of population and housing, it has been conducted every 10 years (in years that end in zero) since. Originally established as a means of determining the number of residents in each state for purposes of allocation of Congressional seats, the content of the census was expanded over time to include hundreds of additional data elements.

Census of Population and Housing. For decades the decennial census included a short form administered to every household and a long form administered to one in six households. The short-form elicited basic information while the long-form collected data on a large and ever-growing number of variables. For the 2010 census the long form was discontinued and only an abbreviated short form was administered. The American Community Survey (ACS), described below, was instituted in 2005 to replace the long form. Data collected via the ACS was used for the first time to supplement data collected by the census in 2010. Box 6.1 provides an overview of the 2020 census and discusses some of the issues surrounding its administration.

By definition, a census includes a complete count of the population. However, it is increasingly difficult to strictly apply this term to the US decennial census. While the census ostensibly counts every resident, it falls short of a true census in two respects. First, every decade a certain segment of the population is missed in the enumeration resulting in some level of undercount. While the undercount is typically less than 3%, its mere existence creates myriad problems. This undercount tends to be concentrated among certain segments of the population,

**Box 6.1: The 2020 Census of Population and Housing**

In 2020, the US Census Bureau administered the twenty-fourth decennial census. The 2020 data collection effort involved the mailing of questionnaires to every known household. Each household received a form with the ten core questions (the short form). Prior to the mailing of the questionnaires, postcards were mailed to every household to alert residents to the coming survey instrument. Included with the questionnaire were instructions for completing the form and an envelope for return mail. In addition to the mail-in option, Americans were given the option of completing the census questionnaire on line via the internet. Other forms of the questionnaire are used for individuals with non-household living arrangements such as those living in group quarters. (The official date for the census count is April 1 of the census year. Since not all responses can be submitted on that date, census forms request information on the respondents status as of April 1.)

The core questions in the 2020 census related to the information required for political redistricting purposes. These questions captured data on the age, sex, race/ethnicity and relationships of household members. This information was subsequently used nationally for the allocation of seats in the US House of Representatives and, at the state level, for adjusting the boundaries of legislative districts and drawing boundaries for new districts as appropriate. Additional details on the 2020 census and its implementation are available at www.census.gov.

The implementation of the decennial census is always a challenge, and the 2020 administration faced unprecedented issues. For starters, the 2020 census was considered to be underfunded and, even then, the funds that were allocated were slow in being released. In the months leading up to the administration of the census the US was experiencing higher than usual employment levels. This created a challenge when it came to hiring the 500,000 temporary workers that the project required.

Further, in March of 2020 the COVID-19 pandemic struck the American population, severely limiting the data collection activities of the Bureau. The pandemic also disrupted traditional data collection activities in that access to some group quarters was limited. College dormitories were closed and off-campus students sent home, thereby skewing data on the distribution of college students as of the census date of April 1. On top of all of this, the Trump administration attempted to interfere with the census and demanded changes in procedures and data collection. Finally, the data collection period was shortened for political reasons, thereby affecting the completeness of the canvassing. These combined factors resulted in many demographers considering this the weakest census ever conducted.

resulting in overrepresentation of some groups and underrepresentation of others. This fact has important implications, since the results of the census are used as the basis for delineating Congressional districts and allocating government funds. Because of the undercount, the initial release of census data every ten years produces a spate of lawsuits questioning the accuracy of the census. The controversy surrounding the undercount has only become more intense in view of the large number of undocumented residents entering the US in recent years.

***The American Community Survey was instituted to supplement the decennial census with data collection on an annual basis.***

The second factor diminishing the enumeration's value as a census is the discontinuation of the long form. Much of the data historically collected through the long form was now obtained via the American Community Survey. The small, targeted sample size (one in 20 households) means that detailed data are not as likely to be available for small geographic areas and that a large margin of error is generated. While the use of sampling significantly reduces the cost associated with conducting the census, the ACS generates figures that some might assume (incorrectly) to represent complete counts. The trade-off is that the ACS assures a steady flow of data between decennial censuses.

Census data are available for virtually every formally designated geographic unit in the United States. Statistics generated by the census are disseminated for states, counties, ZIP Codes, metropolitan areas, and cities. Statistics are also produced for specially designated areas created by the Census Bureau, including blocks, census tracts, block groups, and block numbering areas.

Very little of the results of the census are presented in print form today, although detailed data can be readily accessed via the internet. The Census Bureau makes certain databases—referred to as summary tape files (STFs)—available to those who want to further analyze the data. These databases do not include raw data (i.e., individual records) from the census but preselected aggregations of data. Public use microdata samples (PUMS) do include raw data and are available from the 2020 census, stripped of any information that would identify individual respondents. PUMS files involve a sample of records from areas containing at least 100,000 persons.

***Many private data vendors acquire census data and repackage it for sale to the public.***

After the 1980 census, many private data vendors began to acquire census data and repackage it for sale to the public. In fact, joint public–private projects were involved in converting census data to the ZIP Code level, a geographic unit with a great deal of utility for the business community. Private sector marketing of census data was even heavier after the 1990 census, with commercial data vendors providing population estimates and projections at the census tract level during the intercensal period. The expansion of private sector exploitation of the wealth of information provided through the census has continued with subsequent censuses.

Economic Census. An economic census was initiated in 1810 and is conducted every five years in years that end in 2 and 7. The economic census involves a complete count of all US business operations. While not containing the detail found in the collection of data on population and housing, the economic census represents the only comprehensive source of data on US business establishments and their characteristics.

The economic census covers businesses engaged in retail trade, wholesale trade, service activities, mineral industries, transportation, construction, manufacturing, and agriculture, as well as government services. The information collected includes data on sales, employment, and payroll, along with other, more specialized data. These data are available for a variety of geographic units, including states, metropolitan areas, counties, and incorporated places of 2,500 or more population.

Every business is classified using the North American Industrial Classification system (NAIC). The NAIC system assigns a code to all businesses, allowing them to be grouped into standard categories for statistical purposes. Aggregated information on businesses within the NAIC categories is available at the county level with distribution primarily through the internet. For example, for a sample US county the 2022 economic census found 1434 constructions companies with 218,288 employees and payrolls of $914 million, 2767 retail trade outlets with 37,339 employees and payrolls of $1.1 billion, and 564 manufacturing facilities with 27,031 employees and payrolls of $1.9 billion.

## Registration Systems

A second method of data collection that generates information for demographers is represented by registration systems. A registration system involves the systematic compilation, recording, and reporting of a set of events, institutions, or individuals. The implied characteristics of a registry include the regular and timely recording of the phenomenon in question. Most registration systems relevant to this discussion are maintained by some branch of government, although other sponsors of registration systems exist as well.

***A registration system involves the systematic compilation, recording, and reporting of a set of events, institutions, or individuals.***

The best known registration activities in the United States are those related to "vital events", such as births and deaths. The most extensive registration systems are maintained by the National Center for Health Statistics (NCHS) and its parent agency, the Centers for Disease Control and Prevention (CDC). Other useful systems are maintained by the Social Security Administration (SSA), the Center for Medicare and Medicaid Services (CMS), and Immigration and Customs Enforcement (ICE). Lists maintained of members by trade groups and professional associations (such as the American Medical Association and the American Hospital Association) are placed in this category because such lists have many of the characteristics of registries.

Data from registration systems maintained by the federal government (and some state and local governments) have become increasingly available due to the Internet. The federal government has made an effort to make access to its data easier, and most of the information is free of charge. Other registries vary in their accessibility, some entities (e.g., American Hospital Association) charge fees for the data, and others (usually private organizations) limit access to their data.

A variation on registration systems increasingly deployed by demographers involves administrative records. Administrative records systems are not necessarily intended to be complete registries of all events, organizations, or individuals but to provide a record of the transactions of those included within that group. Thus, the list of all Medicare beneficiaries (enrollees) would constitute a registry, but the data generated by virtue of the beneficiaries' encounters with the healthcare system would be considered administrative records (since not all beneficiaries would use services during a given time period).

***Administrative records systems provide a record of the transactions of those included within that registration database.***

Administrative records can serve a useful function to the extent that they provide access to sources of data not otherwise available. However, unlike other forms of data generation such as census and surveys, the raw data are not strictly under the control of those who establish the data file. Administrative records may be submitted by a variety of parties, creating inherent problems in data quality and standardization. A great deal of effort is currently being expended to improve the accessibility of data maintained by federal agencies.

### *Vital Statistics*

Vital statistics refers to data on births, deaths, and (increasingly) disease incidence/prevalence. (Statistics on marriages and divorces are also considered "vital statistics" but are not routinely collected with other vital data.) The collection of data on vital events is initially the responsibility of local government (i.e., city or county government). The local health department is the primary collector of birth and death statistics, while the local court clerk's office is responsible for the recording of marriages and divorces. Data collected at the local level are forwarded to the appropriate vital statistics registry within the respective state governments. A state agency compiles the data from the various localities and subsequently transfers the data (in the case of births and deaths) to the National Center for Health Statistics (NCHS). The NCHS has responsibility for compiling and publishing vital statistics for the nation and its various political subdivisions. Box 6.2 provides more information on the National Center for Health Statistics.

***Vital statistics refers to data on births, deaths, and (increasingly) disease incidence/prevalence.***

Fertility Data. A standard birth certificate is used in the United States to collect data on the time and date of birth, place of occurrence and the mother's residence, birth weight, pregnancy complications, mother's pregnancy history, mother's and father's age and race/ethnicity, and mother's education and marital status. Virtually all births are recorded, making birth certificates an invaluable source of data on births and attributes associated with mothers and their babies. Data are collected based on the mother's place of residence as well as the location where the birth occurred. This registry is the source of data on births and serves as the basis for calculating most of the fertility metrics discussed in Chap. 7. Box 6.2 presents the data available from a standard birth certificate.

***Virtually all births are recorded, making birth certificates an invaluable source of data on births and attributes associated with mothers and their babies.***

**Box 6.2: Items Included on the Standard Certificate of Live Birth**

| *Child* | *Mother* | *Father* | *Pregnancy* |
|---|---|---|---|
| Name | Name | Name | Pregnancy history |
| Sex | Age | Age | Date of last normal menses |
| Date of birth | State of birth | State of birth | Month prenatal care began |
| Hospital/name | Place of residence | Race/ethnicity | Prenatal visits |
| County of birth | Race/ethnicity | Education | Pregnancy complications |
| Birth weight | Marital status | Relation to child | Concurrent illnesses |
| Apgar score | Education | | Congenital anomalies |
| | | | Method of delivery |
| | | | Medical risk factors |
| | | | Obstetric procedures |

*Source* National Center for Health Statistics

Mortality Data. A standard death certificate is completed for virtually every death that occurs in the United States, making for a very complete record of mortality data. Information gathered on the death certificate includes age, race/ethnicity, sex, place of residence, usual occupation, and industry of the decedent, along with the location where the death took place. In addition, data are collected on the immediate and secondary causes of death, as well as on any other significant conditions. A separate certificate is used for fetal deaths.

The data collected on the standard death certificate include primary cause of death, contributing causes, and individual demographic and socioeconomic characteristics such as sex, race, ethnicity, last occupation, place of residence, and place of death. Using these data, demographers can begin to study the relationship between the cause of death and a variety of demographic variables.

Death statistics are available in government publications and increasingly electronically via the internet. The compiled statistics are typically presented for both the place of occurrence of the vital event (e.g., the location of the hospital) and the place of residence of the effected individual. Considerable detail is provided by the NCHS for a wide range of geographic units including states, metropolitan statistical areas (MSAs), counties, and urban places. Data for other geographic areas may be available through state and local governmental agencies. Yearly summary reports are produced and published by the National Center for Health Statistics, and periodic updates are available through the monthly vital statistics reports. Local and state health departments are increasingly making birth and death statistics available on-line.

While death certificates represent a significant source of data for mortality analysis, there are issues that require caution in their use. There is no universal agreement as to the determination of which factor is the immediate cause of death. There are, in fact, differences that exist from community to community with regard to the classification of contributing and proximate factors. Issues related to the correct allocation of cause of death were illuminated as a result of the COVID-19 pandemic.

There may also be a tendency, hopefully not widespread, to misrepresent the cause of death for various reasons. There may be reluctance, for example, to specify AIDS or some other sexually transmitted disease as a cause of death. Similarly, there may be reticence with regard to specifying alcohol- or drug-related conditions as the cause of death. The slippage with regard to accurate classification of cause of death is also exacerbated due to the trend toward employment of coroners who are not physicians. In fact, in some jurisdictions, the coroner may be an elected office. For these reasons, it is important to use mortality data with caution and certainly to consider the full variety of contributors to mortality.

***Birth and death statistics are available in government publications and increasingly via the Internet.***

Birth and death statistics are available in government publications and increasingly via the internet. The compiled statistics are typically presented for both the place of occurrence of the vital event (e.g., the location of the hospital) and the place of residence of the effected individual. Considerable detail is provided by the NCHS for a wide range of geographic units including states, metropolitan statistical areas (MSAs), counties, and urban places. Data for other geographic areas may be available through state and local governmental agencies. Box 6.3 describes the activities of the National Center for Health Statistics.

**Box 6.3: The National Center for Health Statistics**
The National Center for Health Statistics (NCHS) is considered by many to be the Census Bureau of healthcare. As a division of the Centers for Disease Control and Prevention (CDC), the NCHS performs a number of invaluable functions related to data on health and healthcare. Since 1960, the Center has carried out the tasks of data collection and analysis, data dissemination, and the development of methodologies for research on health issues. The NCHS also coordinates the various state centers for health statistics.

One of the center's responsibilities includes the compilation, analysis, and publication of vital statistics for the United States and each relevant subarea. This is a massive task, and the results provide the basis for the calculation of fertility, mortality, marriage, and divorce rates. These statistics, in turn, provide the basis for various population estimates and projections made by other organizations. The compilation and analysis of data on morbidity is another important function, and the Center has been responsible for the development of much of the epidemiologic data available, for example, on chronic disease and AIDS.

In addition to the data compiled from various registration sources, the Center is the foremost administrator of healthcare surveys in the nation. Its sample surveys are generally large scale and fall into two categories: community-based surveys and facility-based surveys. The former involves the collection of data from a sample of community residents; the latter involves the collection of data from institutions or organizations such as doctors' offices, nursing homes and hospitals.

The data collected through NCHS programs are disseminated in a variety of ways. Much of the information is disseminated as printed material. The Center's publications include annual "books" such as *Health, United States* (the "official" government compendium of statistics on the nation's health), and series of publications such as *Vital and Health Statistics*. These publications are available today primarily via the internet at the NCHS website. The NCHS also sponsors conferences and workshops offering not only the findings from Center's research but training in its research methodologies.

Morbidity Data. The study of morbidity has become a growing focus of demographers for a number of reasons. These include the growing emphasis on population health, the declining significance of mortality, and the shift from an emphasis on acute conditions to chronic conditions. The interest in morbidity on the part of demographers, epidemiologists, health planners and medical scientists steadily increased over the last quarter of the twentieth century as the connection between demographic variables and morbidity differentials became clearer.

***The study of morbidity has become a growing focus of demographers due to the declining impact of mortality on the US population.***

Since there is no national repository of morbid diseases, much of what we know about various diseases is drawn from disease registries. The Centers for Disease Control and Prevention (CDC) and its affiliated organization, the National Center for Health Statistics, maintain registries for certain conditions. The most long-standing registry is the "reportable disease" database maintained by the CDC. This registry tracks dozens of communicable diseases whose must be reported by law. This registry has become less important over time due to the diminished importance of infectious diseases. However, its usefulness has been on display recently as it has alerted public health officials to the emergence or re-emergence of certain infectious diseases that we had thought to be eliminated.

There are other registries maintained by these federal agencies that track non-communicable diseases. There are registries for birth defects, poisoning, and various types of cancer, to name a few. Data based on these registries are typically made available to the public through various publications and increasingly on-line. (The other major source of morbidity data is sample survey, discussed below.)

***Since there is no national repository of morbid diseases, much of what we know about various diseases is drawn from disease registries.***

Immigration Data. Data on immigration patterns and the characteristics of immigrants historically have been of interest to demographers because of the implications of these phenomena for population change. Today, however, data on immigration are of increasing interest due to the growth of illegal immigration and concerns over the impact of immigration on other demographic processes and attributes. Monitoring international migration is a responsibility of the federal government and the agency responsible for monitoring and reporting on immigration trends is the Immigration and Customs Enforcement (ICE) agency, formerly the Immigration and Naturalization Service (INS), now within the Department of Homeland Security.

***Monitoring international migration is a responsibility of the federal government and the Immigration and Customs Enforcement (ICE) agency has primary responsibility.***

Data are collected related to legalization applications, refugees, asylum applicants, nonimmigrant entries, naturalizations, and enforcement and made available by means of published reports and the internet. Data on visas issued is maintained by the Department of State and data on immigrant visas are available on everyone legally entering the United States. The information collected includes port of admission, country of birth, last residence, nationality, age, sex, occupation, and the ZIP Code of the immigrant's intended residence. The factors that contribute to the level of immigration are discussed in Chap. 9.

Because of the increase in illegal immigration, a growing amount of data is generated as a result of border monitoring and internal police activities. Annual estimates of illegal immigration are generated by ICE.

Data on immigration are made available through yearly statistical summaries, more frequent shorter reports, and via the internet. While the published reports contain data for states and MSAs, tabulations by county and ZIP Code are possible by accessing ICE data files.

## Surveys

A sample survey involves the administration of an interview form to a portion of a target population that has been systematically selected. The sample is designed so that the respondents are representative of the population being examined. This allows conclusions to be drawn for the total population based on the data collected from the sample.

The use of sample surveys has several advantages relative to the census and registry methods. Two of the major advantages are more frequent data collection and the ability to probe more deeply into the subject under study. The relatively small sample sizes for such surveys have the additional advantages of quicker turnaround time and easier manipulation than large-scale operations such as the census.

***The use of sample surveys has several advantages relative to the census and registry methods.***

On the other hand, surveys have their disadvantages. Since they involve a sample, there is some slippage in accuracy relative to censuses. Perhaps the most serious shortcoming related to demographic analysis in terms of the inability to compile adequate data for small geographic units due to small sample sizes.

The federal government is the major source of survey data related to variety of topics. Through various agencies, the federal government administers a number of ongoing surveys that involve information of interest to demographers. The National Institutes of Health and the Centers for Disease Control and Prevention conduct surveys that generate data useful to health demographers. The Census Bureau conducts surveys related to population characteristics, housing and economic trends. The Department of Labor conducts surveys related to employment, occupational trends and so forth. Some of the more useful federal surveys are discussed in Box 6.4.

**Box 6.4: Representative Surveys Sponsored by the Federal Government**
The combined agencies of the federal government represent the nation's largest data collection force. The entries below describe a sample of the federal survey activities that have particular relevance for demographers.

The *National Health Interview Survey (NHIS)* is an ongoing national survey of the US non-institutionalized civilian population. Each year, a multistage probability sample of between 36,000 and 46,000 households (92,000–135,000 persons) is generated for inclusion and interview. The data gathered are quite detailed and the NHIS contains a number of demographic items. Demographic information collected includes age, race, sex, marital status, occupation, and income. Health questions relate to physician visits, hospital stays, restricted-activity days, long-term activity limitation, and chronic conditions. For some surveys, questions regarding AIDS knowledge and attitudes have been added to the survey. Food nutrition knowledge, smoking and other tobacco use, cancer, and polio are also subjects sometimes addressed.

The *National Hospital Discharge Survey (NHDS)* is a continuous nationwide survey of inpatient utilization of short stay hospitals. All hospitals with six or more beds reporting an average length of stay of less than 30 days are included in the sampling frame. A multistage probability sampling frame is used to select hospitals from the National Master Facility Inventory and discharge records from each of the hospitals. The sample has ranged from 192,000 to 232,000 discharge records each year from 1970 to the present. The information collected includes demographic, clinical, and financial characteristics of patients discharged from short-stay hospitals. The variables abstracted range from the age, race, sex, and marital status of the person discharged to source of payment, discharge status, diagnosis, and length of stay. Hospital information on bed size and type of ownership is also gathered.

The *National Ambulatory Medical Care Survey (NAMCS)* is a nationwide survey designed to provide information about the provision and utilization of ambulatory healthcare services. The target universe includes office visits made by ambulatory patients to physicians engaged in office practice. A multistage probability sampling frame is used to select physicians from the master files maintained by the American Medical Association and the American Osteopathic Association. Their records for a sample of office visits during a randomly assigned one-week period are then examined. Recent samples contain about 70,000 records. Data regarding the age, race, ethnicity, and sex of the patient are gathered along with the reason for the visit, expected source(s) of payment, principal diagnosis, diagnostic services provided, and disposition of the visit.

The *Medical Expenditures Panel Survey (MEPS)* gathers information on health status, access and use of medical services, associated charges, and sources of payment as well as insurance coverage for a sample of patients. Cosponsored by the Agency for Healthcare Research and Quality (AHRQ) and NCHS, MEPS is designed to generate data on the types of health services Americans use, the frequency with which they use them, the amount paid for these services, and the source of payment. In addition, MEPS provides information on health insurance coverage.

The *Current Population Survey (CPS)* is the Census Bureau's initiative for gathering detailed demographic information between decennial censuses particularly as related to the labor force. This monthly survey of households conducted by the Bureau of Census for the Bureau of Labor Statistics provides a comprehensive set of data on the labor force, employment, unemployment, persons not in the labor force, hours of work, earnings, and other demographic and labor force characteristics. Since 1960, the sample size has ranged from 33,500 to 65,500 households. Of particular interest to the health care industry are the data on the extent and type of health insurance coverage for the total US population.

The *National Survey of Family Growth (NSFG)* is a periodic survey conducted to gather information on fertility, family planning, and aspects of maternal and child health that are closely related to childbearing. It is a multi-stage probability sample of civilian non-institutionalized women aged 15–44. A wide range of information is collected through personal interviews including the age, race, marital status, education, income, and religion of each woman surveyed. Family planning data collected range from age at first contraceptive visit and contraceptive use to sources of family planning services and sources of payment for such services.

The *Behavioral Risk Factor Surveillance System (BRFSS)*, sponsored by the centers for Disease Control and Prevention was initiated in 1995 to collect information on the health behavior and lifestyles of the US population. Over 150,000 persons respond to the survey annually. Data are collected on such timely items as smoking patterns, alcohol and drug use, seat belt use, and obesity, as well as other factors that might contribute to one's health risk profile. States and localities have the option of sponsoring additional behavioral risk research through expanding the number of surveys conducted in their area or adding questions related to topics of interest to the community.

As previously noted, the American Community Survey (ACS) has been introduced by the Census Bureau as a replacement for the long-form data collection instrument for the decennial census. Increasingly, demographers and others have come to rely on data from the ACS for a wide variety of uses despite the shortcomings of this survey. Box 6.5 presents information on the ACS.

A number of private organizations also conduct surveys of interest to demographers. There are research institutes and "think tanks" that collect data on reproductive practices, immigrant characteristics, consumer behavior patterns and so forth. The results of these surveys are typically not available for levels of geography below the nation, although some may generate state-level data. Demographic-oriented organizations like the Population Reference Bureau may also conduct surveys, although they typically repackage data from other sources.

Commercial data vendors also conduct surveys that contain data useful to demographers. At least two vendors conduct national surveys annually on health-related characteristics and health behavior. Other data vendors may extract health-related data from national syndicated surveys and package this information with their demographic data. Some of these data sets are considered proprietary and generally are only available to established clients. Other data may be available for sale to the public.

**Box 6.5: The American Community Survey**

The American Community Survey (ACS) is an ongoing survey that provides data every year most US communities. The ACS includes questions related to around 30 population topics and 25 housing topics. The survey was introduced in 2005 for annual administration, and in 2021 data were obtained from nearly 2,000,000 households. People selected as part of the sample are required by law to respond.

Unlike the decennial census, the ACS involves continuous measurement of the topics under study. Demand for current, nationally consistent data from a wide variety of users led federal government policymakers to consider the feasibility of collecting social, economic, and housing data continuously throughout the decade. The benefits of providing current data, along with the anticipated decennial census benefits in cost savings, planning, improved census coverage, and more efficient operations, led the Census Bureau to implement the American Community Survey (ACS).

The following criteria were considered important for an effective on-going survey:

- Data would be collected continuously by using independent monthly samples.
- Three modes of data collection would be used: mail-out, telephone non-response follow-up, and personal visit non-response follow-up.
- The survey reference date for establishing housing unit occupancy status, and for many characteristics would be the day the data were collected. Certain data items would refer to a longer reference period (for example, "last week," or "past 12 months").
- The survey's estimates would be controlled to intercensal population and housing estimates.
- All estimates would be produced by aggregating data collected in the monthly surveys over a period of time so that they would be reported annually based on the calendar year.

Data generated by the ACS are presented for various levels of census geography, with the most granular level being the census block group. Since the sample is not large enough to produce accurate estimates for all geographies in any particular year, the results of the ACS are published in three temporal versions: 1-year data, combined 3-year data, and combined 5-year data. The more years that are combined the greater the sample size and the more reliable the estimates. For larger geographies, one year of data may suffice but more often than not a smaller community or lower level of geography will necessitate the combining of years. Combined data, of course, have the disadvantage of representing different time periods, sometimes combining data separated by four years in time.

The ACS does not have the statistical power of the one-in-six household long form used by the Census Bureau in the past and demographic purists raise some issues with the methodology. However, the benefit of having continuous data collection is thought to outweigh any drawbacks. The most direct way to access data from the American Community Survey is through the Census Bureau website accessed at www.data.census.gov.

## Synthetic Data

*Synthetic data* refers to statistics that are produced in the absence of actual data using models that simulate reality. Synthetic data are generated by merging existing demographic data with assumptions about population change to produce estimates, projections, and forecasts. The outputs generated through synthetic data calculations are particularly valuable given that census and survey activities are constricted because of budgeting and time considerations. Consequently, there is a large and growing demand for information between the years when data are actually collected. This demand is being met by government agencies and commercial data vendors.

***Synthetic data refers to statistics that are produced in the absence of actual data using models that simulate reality.***

### *Population Estimates and Projections*

Demographers have long used population estimates and projections in the absence of actual data, and a variety of techniques are utilized to generate these figures. Population estimates for states, MSAs, and counties are prepared each year as a joint effort of the Census Bureau and the designated state agency under the Federal-State Program for Local Population Estimates (FSCPE). The purpose of the program is to standardize data and procedures so that the best quality estimates can be derived.

Most states also generate population estimates and projections that are available through state agencies. However, these figures are often produced at irregular intervals, and thus may be quite dated. The reader is also encouraged to evaluate the quality of these data to the best of his or her ability. For additional information on population projections and estimates see Smith et al., (2000).

Population estimates and projections generated by government agencies have historically been the only ones available. Today, however, a number of data vendors provide these figures. These vendor-generated data are often made available down to small units of geography (e.g., the census tract) and in greater detail (e.g., sex and age breakdowns) than government-produced figures. They offer the ability to generate estimates and projections for "custom" geographies (e.g., a market area) not available for government-generated statistics. The drawback, of course, is that some precision is lost as one develops calculations for lower levels of geography and for population components. Furthermore, the methodologies for the generation of synthetic data requires assumptions about the target population that may be faulty. Nevertheless, the ease of accessibility and timeliness of these vendor-generated figures have made them a mainstay for health planners and researchers.

Issues have been raised concerning the quality of the synthetic data produced by both government agencies and commercial data vendors (Cropper et al., 2011). Data users typically need the latest information possible, and in an effort to be expedient the question of quality sometimes is a secondary concern. Any evaluation of the quality of synthetic data requires an understanding of the recency and quality of the historical data being used as a basis for the estimates and projections.

Occupational Projections. The Bureau of Labor Statistics (within the US Department of Labor) maintains data on all occupational categories within the economy, including healthcare occupations. As part of the Bureau's responsibilities, it produces projections on the size of various occupational categories in the United States for 10–15 years into the future. Six models are used to generate projections, each containing a number of variables reflecting different scenarios related to changes in the total labor force, the aggregate economy, industry demand, and industry employment, among other factors. Three sets of employment projections are generated based upon differing sets of assumptions. Of interest here are the various categories of clinical occupations (e.g., physicians, dentists) and nonclinical healthcare occupations (e.g., insurance claims managers, medical records personnel). In recent years, health-related occupations have been prominent among the occupations with the greatest projected growth.

### *Demographic Forecasts*

Forecasts resemble projections in that they offer an estimation of conditions for some future point in time. Forecasts, depending on the methodology used, may or may not rely heavily on historical trends. The distinguishing characteristic of forecasts is that they attempt to take likely future developments into consideration. For this reason, there is more of a subjective dimension to the formulation of forecasts relative to projections. The forecaster is likely to try to anticipate future developments that will affect the size of the population, perhaps quite independent of any prior developments. Econometric models are often used to generate forecasts rather than projections because of their tendency to incorporate a number of variables that are thought to reflect future developments.

Projections for various occupational categories are available from the Department of Labor through regularly published reports. The percentage distribution for the labor force (matrix coefficients) can also be obtained and these are sometimes used by other organizations to produce subnational occupational projections.

## Sources of Data for Demographic Analysis

There are numerous sources of data for demography available today and the number of sources continues to grow. The sections below group these sources into four main categories: government agencies, professional associations, private organizations, and commercial data vendors.

It should be noted that the "products" available from these sources fall into two categories: (1) reports that summarize the data and (2) the actual data sets themselves. Historically, data access was limited to summary tables provided by the organization, agency or vendor. Today, however, there is a trend toward providing the entire data set for use by researchers, analysts, and other data users. In reviewing the sources that follow, this distinction in format should be kept in mind. Box 6.6 specifies sources of specific categories of data.

### *Government Agencies*

Governments at all levels are involved in the generation, compilation, manipulation and/or dissemination of data on compositional attributes. The federal government, through the decennial census and related activities, is the world's largest processor of demographic data. The Census Bureau remains a major source of relevant demographic data, and other federal agencies are custodians of data for such related topics as fertility, mortality, income, employment and occupations, and migration.

State and local governments are also sources of data on compositional variables. State governments generate a certain amount of demographic data, with each state having a state data center for demographic projections. Vital statistics data can often be obtained in the most timely fashion at the state level, in fact. States vary, however, in the types and quality of data they generate. University data centers may also be involved in the processing of demographic data. Local governments may generate demographic data for use in various planning functions. City and county governments may produce population projections, while county health departments are responsible for the collection and dissemination of vital statistics data.

***Governments at all levels are involved in the generation, compilation, manipulation and/or dissemination of data on compositional attributes.***

### *Professional Associations*

In recent years, many professional associations have made an increasing amount of information on their members available to the research and business communities. Not only do such organizations have an interest in exchanging information with related groups, but they also have recognized the revenue generation potential of such databases. Some of the databases provided by professional associations include only basic information, while others offer a wealth of detail.

### *Private Organizations*

Many private organizations (mostly not-for-profit) collect and/or disseminate demographic data. Voluntary associations often compile, repackage and/or disseminate such data. The American Cancer Society, for example, distributes morbidity and mortality data as it relates to its areas of interest. Some organizations, like Planned Parenthood, may commission special studies on fertility or related issues and subsequently publish this information.

Many organizations repackage data collected elsewhere (e.g., from the Census Bureau or the National Center for Health Statistics) and present it within a specialized context. The Population Reference Bureau, a private not-for-profit organization, distributes population statistics in various forms, for example. Some, like the American Association of Retired Persons (AARP), not only compile and disseminate secondary data but are actively involved in primary data collection.

### *Commercial Data Vendors*

Commercial data vendors represent a fourth category of sources of demographic databases. These organizations have emerged to fill perceived gaps in the availability of various categories of data. These include commercial data vendors that establish and maintain their own proprietary databases, as well as those that reprocess and/or repackage existing data. For example, there are vendors that maintain databases that make this information available in a variety of forms. Major data vendors that do not necessarily create databases may incorporate demographic data into their business database systems. Some data vendors conduct major nationwide health consumer surveys. Box 6.6 identifies some of the federal sources of demographic data.

**Box 6.6: Selected Federal Sources of Demographic Data**

| *Information category* | *Source* |
|---|---|
| *Population data* | |
| Size | ACS, census, CPS, vendors |
| Characteristics | ACS, census, CPS, vendors |
| Estimates and projections | CB, CPS, vendors |
| *Vital statistics* | |
| Births | NCHS |
| Deaths | NCHS |
| Disease incidence/prevalence | NCHS |
| Abortions | NCHS |
| Marriages/divorces | CB |
| *Migration data* | |
| Internal migration | ACS, CB, CPS, IRS |
| Immigration | ICE |
| Education data | USDOE |
| Income/poverty data | CB, CPS |
| Labor force data | CB, BLS |
| Food/nutrition data | USDA |
| Environmental data | EPA |
| Criminal justice data | USDOJ |

*Legend* ACS = American Community Survey
BLS = Bureau of Labor Statistics
CB = Census Bureau
CPS = Current Population Survey
EPA = Environmental Protection Agency
ICE = Immigration and Customs Enforcement
IRS = Internal Revenue Service
NCHS = National Center for Health Statistics
USDA = US Department of Agriculture
USDOE = US Department of Education
USDOJ = US Department of Justice.

## Key Points

- Access to data is critical for the demographic enterprise, since information on population characteristics are utilized for virtually every analysis.
- Although there is not a single source of demographic data, there are a variety of options to choose from.
- Demographic data are generated by means of censuses, sample surveys, and registries and increasingly, generators of synthetic data.

- For decades demographers have relied on the data generated through the decennial census as the "gold standard" of demographic data.
- There are various issues with the decennial census that have made it less valuable as a source of demographic information.
- The American Community Survey was initiated to supplement the decennial census.
- The use of sample surveys has several advantages relative to the census and registry methods.
- Synthetic data are generated that include estimates, projections and forecasts.
- Government agencies (at all levels) are sources of demographic data.
- The federal government is the largest producer of demographic data in the world.
- Other sources of demographic data include private organizations, associations, and commercial data vendors.
- Data vendors have emerged to repackage existing demographic and produce their own
- Monitoring international migration is a responsibility of the federal government and the Immigration and Customs Enforcement (ICE) agency has primary responsibility.

## Discussion Questions

6.1 What are the drawbacks of a decennial census and how has the American Community Survey served to address some of those deficiencies?

6.2 Every census produces an undercount that typically is less than 3%. However, this undercount affects some segments of the population more than others. What are the implications of an undercount of minorities, for example, when it comes to redistricting for Congress and the distribution of federal funds?

## Exercises

*Exercise 6.1: Exploring Federal Sources of Data*

For this exercise, students should surf the web for the various federal agencies that are likely to hold data of use to demographers. Students should start with the Department of Commerce website (which includes the Census Bureau), the Department of Health and Human Services website (which includes the Center for Disease Control and Prevention), the Department of Housing and Urban Development, and so forth. The results of this exercise should be a fairly lengthy list of sources of data of use to demographers.

*Exercise 6.2: PolicyMap Search*

For this exercise, students should access policymap.com to determine the types of data available via this extensive on-line database. Consider the various attributes that we have been learning about and determine the extent to which those data can be found at policy map. Explore the mapping options and be able to report to the class information useful to them for their demographic research.

## Reference

Cropper, M., McKibben, J.N., Swanson, D.A., & Tayman, J. (2011). *Vendor accuracy study: 2010 estimates versus census 2010*. ESRI.

### Additional Resources

American Hospital Association (Annual). *AHA guide*.
American Medical Association (Annual). *Physician characteristics and distribution in the US*.
Bureau of Labor Statistics. www.bls.gov
Census Bureau. www.census.gov
Centers for Disease Control and Prevention. www.cdc.gov
Health Resources and Services Administration. www.hrsa.dhhs.gov
Immigration and Customs Enforcement. www.ice.gov
McCullough, M. (2023). Policy map: Easy on-line mapping for business users. *CIO Review*. https://www.cioreview.com/policymap
National Center for Health Statistics. www.cdc.gov/nchs
Smith, S. K., Tayman, J., & Swanson, D. A. (2000). *State and local population projections: methodology and analysis*. Plenum.

CHAPTER 7

# Demographic Processes: Fertility

**Abstract** This chapter addresses the first of the three basic demographic processes: fertility, mortality and migration. The importance of fertility to demographers and to society is recognized and the various ways of assessing fertility trends reviewed. Students are exposed to the various calculations that are utilized to measure the fertility attributes of a population, and trends in fertility patterns for the US and the world are described. Issues surrounding the reproductive process are discussed and their relevance for society reviewed.

## Introduction

Fertility is a social process requiring the biological interaction of two persons within a defined economic, social and/or political context. The reproductive experience of a population involves all factors related to sexual behavior, pregnancy, and birth outcomes. The number of births as well as the characteristics of those births, along with factors describing the characteristics of mothers, fathers and newborns, constitute the basis for fertility analysis. Fertility behavior is viewed broadly here and includes pre-pregnancy behavior, prenatal care, health-related activities during pregnancy, and pregnancy outcomes. This perspective involves the interaction of culture, technology and economic conditions with fertility behavior.

***Fertility is a social process requiring the biological interaction of two persons within a defined economic, social and/or political context.***

Fertility plays an important role in shaping the demographic makeup of a population. The level of fertility, along with an area's mortality and migration characteristics, determines the size and composition of any population.



R. K. Thomas, *Demography: An Introduction to Population Studies*, Springer Texts in Social Sciences, https://doi.org/10.1007/978-3-031-56623-3_7

Longer-term fluctuations in the number of births translate into changes in the size of age cohorts over time. Reproductive behavior today will have implications for society tomorrow. The continued trend toward smaller households and families as well as the reduction in the proportion of persons living in families means that there will be other as yet undiscernible changes in American society.

The study of fertility is of particular importance to demographers. Along with mortality and migration, it represents one of the three "legs" of the population change "stool". Fertility patterns, along with in-migration, serve to determine the number of individuals added to a population. Given the significant elasticity in the level of fertility, the study of reproductive patterns is an important activity for demographers.

## Concepts and Measures

A variety of concepts are used to describe and measure the fertility process. The starting point is typically the number of births that occur within a population and the fertility rate it represents.[1] The various measures of fertility used by demographers are described below.

***Fertility is measured in terms of the number of births that occur within a population and the fertility rate it represents.***

The level of fertility is typically expressed in terms of rates. The calculation of fertility rates facilitates the comparison of fertility levels across areas that differ in size and/or other characteristics. Comparing the number of births for two cities with populations of 100,000 and 1,000,000, respectively, makes little sense given that the base population producing births is 10 times larger in the latter city. The use of rates also facilitates the identification of trends in the level of fertility over time.

The *crude birth rate* (CBR) is the most basic measure of fertility. It is calculated by dividing the total number of births for a given year (or the average over three years) by the midyear total population for that year (or the midyear in the range if a three-year average of births is taken). This quotient is then expressed as the number of births per 1,000 population. The crude birth rate for the US was 23.7 births per 1000 persons in 1960 and had dwindled to half that rate (11.9) by 2020 (National Center for Health Statistics, 2022). The 2020 birth rate was the lowest ever recorded for the US population. (The formulas for calculating this and other measures of fertility are presented in Box 7.1.)

[1] "Birth rate" is an umbrella term for all measures related to the reproduction process. "Fertility rate" refers to specific birth-related measures. The two terms are sometimes used interchangeably.

***The crude birth rate is the most basic measure of fertility and is calculated by dividing the total number of births for a given year by the midyear population.***

The CBR is adequate for making a general assessment of a population's fertility level and has the advantage of requiring only two pieces of information. However, the denominator for the crude birth rate includes people who are not at-risk of having a birth. Males, very young females, sterile females, and females beyond menopause are not at-risk of giving birth, yet they appear in the denominator of the rate. Fertility rates are greatly affected by age composition, particularly for women, and the CBR cannot account for this. Two populations of the same size could easily have dissimilar CBRs simply because females in the childbearing ages accounted for 20% of one population but 35% of the other. As a result of these shortcomings, more refined measures of fertility have been developed.

The *general fertility rate* (GFR), sometimes referred to simply as the *fertility rate*, represents a refinement of the CBR. It adjusts the denominator of the rate by focusing on the *population at risk* for pregnancy. It is expressed in terms of births per 1000 females aged 15–44 (or 15–49). In 1960, the GFR was 118 births per 1000 women aged 15–44, and by the mid-1990s it had declined to less than 60. The GFR reached a "modern" high of 69 in 2007 but has been declining since then, standing at 55.8 in 2020 (National Center for Health Statistics, 2022).

While the GFR expresses fertility in terms of births per 1000 women in the at-risk age group, it provides no information on fertility for specific age intervals (e.g., women aged 15–19). Additional information can be provided by calculating *age-specific birth rates*.

***The general fertility rate represents a refinement of the crude birth rate, since it uses the population at risk for pregnancy as the denominator.***

**Box 7.1: Calculating Fertility Rates**

Fertility rates are relatively easy to calculate, and in most instances the required data are readily available. Birth data (numerators) are available from vital statistics registries, and population figures (denominators) can be drawn from Census Bureau counts or estimates generated by other sources. These basic rates can be adjusted to reflect other factors such as age and marital status as desired.

$$\text{Crude birth rate (CBR)} = \frac{\text{Number of births in year X}}{\text{Population at midpoint (July 1) in year X}} \times 1000$$

$$\text{General fertility rate (GFR)} = \frac{\text{Number of births in year X}}{\text{Number of women age 15 to 44 (or 49) at midpoint (July 1) in year X}} \times 1000$$

$$\text{Age-specific fertility rate (ASDFR)} = \frac{\text{Number of births in year X to women age y to y + n}}{\text{Number of women age y to y + n at midpoint(July 1) in year X}} \times 1000$$

$$\text{Total fertility rate(TFR)} = \frac{\text{Sum of ASFRs} \times 5}{1000}$$

$$\text{Gross reproduction rate (GRR)} = \sum_{i=1}^{n} \frac{B_x}{P_x}$$

$$\text{Net reproduction rate(NRR)} = \sum_{i=1}^{n} \frac{B_x}{P_x} - \frac{L_x}{1_0}$$

where $B_x/P_x$ is the female age-specific birth rate for age $x$ to $x + n$; $L_x/1_0$ is the life table survival rate for that age interval, and $\sum_{i=1}^{n}$ indicates that all the products for each ASFR combination are summed. Given that there are eight categories of ASFRs in Box 7.2, n would be equal to 8.

Demographers typically calculate age-specific fertility rates (ASFR) using five-year age intervals. The age-specific fertility rate for women 20–24 years of age, for example, is derived by dividing the number of births to women who are 20–24 years of age by the number of women in that age cohort. The rate is usually calculated for one year, and fertility is expressed in terms of births per 1000 women age 15–44. As can be seen, there are wide differences in the rates. The 25–29 year age cohort exhibited the highest fertility rate in 2020. The rates for women under age 20 and age 40 and over are much lower than those for women age 25–29. Box 7.2 presents age-specify fertility rates for 2020.

**Box 7.2: Age-Specific Fertility Rates United States 2020**

| *Age group* | *Rate per 1000 women* |
|---|---|
| 10–14 | 0.2 |
| 15–19 | 15.3 |
| 20–24 | 62.8 |
| 25–29 | 115.1 |
| 30–34 | 90.0 |
| 35–39 | 94.8 |
| 40–44 | 11.8 |
| 45–49 | 0.9 |

*Source* National Center for Health Statistics

In 1960 (during the peak of the baby boom) the ASFR for 20- to 24-year-olds was 258 (258 births per 1000 women in this age cohort). By 2020 this rate had declined to 63. Since 1980, however, there has been an increase in the fertility rate for older women.

The *total fertility rate* (TFR) is sometimes utilized as a summary measure for age-specific fertility rates. The TFR reflects the hypothetical completed fertility for a population. Technically, the only way to accurately determine how many children a cohort of young women (e.g., those currently under age 15) will have over their lifetimes is to wait 30 or more years until they have completed their childbearing. Since this is not practical, hypothetical measures that allow an analyst to project the completed fertility of a specified cohort without the long wait have been developed.

The calculation of the TFR assumes that a group of 15-year-old females will experience the same age-specific fertility rates during their lives; e.g., at ages 15–19 the cohort will experience a birth rate of 15.3 births per 1000 women per year. Since the interval 15–19 covers five years, the rate is multiplied by five. Adding up all the ASFRs (multiplied by 5) produces a hypothetical total number of births per 1000 women. The TFR calculation yields figures of 1.8 births per woman for 1987 and 1.67 for 2020, both well below the replacement rate of 2.1.

***The total fertility rate is sometimes used as a summary measure for age-specific fertility rates.***

The *gross reproduction rate* (GRR) adjusts the TFR to include only female births. This adjustment makes intuitive sense since it is only females who can bear children. *Replacement-level fertility*, the number of births required for females to exactly replace themselves, is 2.1 births per woman over a lifetime. The implications of declining fertility for the US population are discussed in a later section.

While the GRR meets the demand for a measure of replacement, it fails to account for the mortality experience of both children and mothers. Therefore, an additional refinement, the *net reproduction rate* (NRR), has been created in order to adjust the measure of replacement by accounting for the deaths to women and female children that are known to occur. Adjusting for mortality results in NRRs that are smaller than GRRs. The factors used to adjust the GRR are derived from observed mortality data and the life tables that are based on these data.

***Replacement-level fertility, the number of births required for females to exactly replace themselves, is 2.1 births per woman over a lifetime.***

## Other Fertility Concepts

Other concepts used by demographers include *fecundity* (the physiological ability to reproduce); *age at menarche* (the onset of menstruation); *menopause* (the end of menstruation); *age at first birth*; and *parity* (the number of children women have already had). Since physiological factors can change in response to the physical and social environments (e.g., the historical reduction in age at menarche in the United States), they are likely to affect fertility levels and the demand for services.

Additional important concepts useful to the health demographer include *pregnancy wantedness* (whether the pregnancy was wanted before, at that time, later, or not at all with respect to when pregnancy occurred) and *marital timing* (whether the pregnancy occurred before or after being married). Wantedness and marital timing have been shown to be related to a host of factors, including behavior before, during, and after pregnancy.

## Reproductive Outcome Measures

Another aspect of interest to demographers is reproductive outcomes measures. The US population continues to suffer from high rate of reproductive health problems relative to other similar countries. The implications of negative outcomes extend far beyond demography and affect health status and have consequences for the social and economic aspects of our society.

Among the issues related to reproductive health outcomes are premature births and low-birth weight deliveries. Premature births are those that occur before 37 weeks of pregnancy are completed. Babies may be born preterm because of spontaneous preterm labor or because there is a medical indication that dictates early caesarean birth. Complications associated with a preterm birth include immature lungs, difficulty regulating body temperature, poor feeding, and slow weight gain. Preterm babies may need longer or more intense nursery care, medications, and sometimes surgery.

***Among the issues related to reproductive health outcomes are premature births and low-birth weight deliveries.***

The United States leads the developed world in terms of premature births. Some 10.5% of babies were born prematurity in the US in 2021, representing an increase over previous years. This amounted to over 380,000 preterm births for that year. There are important differences in rates of prematurity by race and ethnicity. African American mothers are nearly 40% more likely to experience premature births and American Indians somewhat more likely. Prematurity is the largest cause of infant deaths.

Low-birth weight babies are often associated with prematurity. "Low birthweight" is designated when a baby is born weighing less than 5 pounds, 8 oz. Some babies recording low birthweight are healthy, even though they are smaller than the norm. But having a low weight at birth can cause serious health problems for some babies.

The primary cause of low birthweight is prematurity (i.e., being born before 37 weeksgestation). Being born early means a baby has less time in the mother's uterus to grow and gain weight. Much of a baby's weight is gained during the latter part of pregnancy.

Another cause of low birthweight is intrauterine growth restriction (IUGR). This occurs when a baby does not grow well during pregnancy because of problems with the placenta, the mother's health, or the baby's condition. Premature babies that have IUGR are both very small and physically immature.

***The US compares unfavorably to comparable counties when it comes to reproductive outcomes.***

Any baby born prematurely is more likely to be very small. However, there are other factors that can also contribute to the risk of very low birthweight. Low birthweight deliveries are more frequent among African Americans, teen mothers, and unhealthy mothers. Babies of mothers who are exposed to illicit drugs, alcohol, and cigarettes are more likely to have low birthweight. Mothers of lower socioeconomic status are also more likely to have poorer pregnancy nutrition, inadequate prenatal care, and pregnancy complications–all factors that can contribute to low birthweight. The constant stress faced by members of certain groups is also considered a factor in poor pregnancy outcomes.

There are other issues associated with reproduction such as infant mortality and maternal mortality. These outcomes are considered in the next chapter on mortality.

## Factors Affecting Fertility

The level of fertility is highly "elastic", and a number of social factors affect the volume of births for a population. Information on factors influencing the birth process, as well as the demographic characteristics of both mothers and babies, makes it possible to determine the characteristics of future populations. The following sections review some of the factors that influence the fertility rate.

### *Reproductive Behavior*

A model for understanding these factors was developed during the 1950s by Davis and Blake (1956) and is still useful today. The model as presented in Box 7.3 illustrates the factors influencing fertility outcomes. These factors do not act independently of each other, although each category represents a separate stage in the fertility process. That is, intercourse must occur first, followed by conception, and last, by successful gestation. The intercourse variable is operationalized in terms of age at first intercourse, frequency of intercourse, time spent in and out of marriage, and age at first marriage.

**Box 7.3: Factors Affecting Fertility**

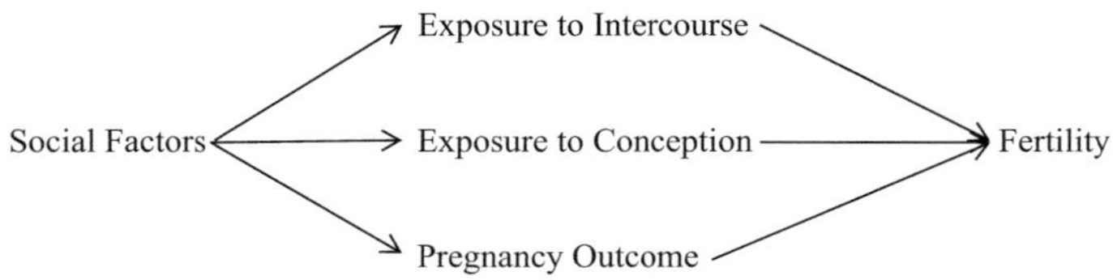


*Source* Davis and Blake (1956)

### *Health Status*

Another consideration is the impact of the health status of the population on fertility levels. It is a biological fact that the ability to reproduce for people (particularly women) who are in very poor health or under considerable stress is reduced. Even people in these circumstances who are able to conceive may face challenges in bringing the pregnancy to term or producing a healthy child. Because of this, careful consideration should be given the health status of the population under study when examining fertility patterns.

### *Compositional Attributes: Biosocial Characteristics*

The age of the mother is an important determinant of pregnancy outcomes. Pregnant women who are very young or over 35 years are considered high-risk (although for different reasons). Earlier initiation of sexual behavior has raised concerns about pregnancies among the youngest women. Very young mothers may not be fully developed themselves and, thus, give rise to more delicate babies. Not only are there associated physical risks for both mother and baby, but babies born under such circumstances start out with social and economic disadvantages. Concerns over births to very young women have been eased somewhat with the reported decline in teenage pregnancies due to greater use of contraceptives and emerging evidence that the health risk for young mothers is not as great as previously believed (Geronimus 2003).

With regard to race and ethnicity, several differences in fertility patterns are apparent. In 2020, the white general fertility rate was 51 births per 1000 women aged 15–44, compared to 54 for African Americans and 56 for Hispanic mothers. Fertility rates for the African American and Hispanic populations are somewhat higher than that for whites, and, as the US population becomes more heavily populated by the two minority groups, the TFR will rise without any real change in fertility. In 2020, the TFR for whites was 1.55 (a historic low), compared to 1.88 for Hispanics, and 1.71 for African Americans.

The greatest value in this information lies in its usefulness for understanding the impact of multiple factors (e.g., age and race) on fertility and fertility-related health services while accounting for compositional changes over time. For example, given racial/ethnic differences in fertility and the younger age structures of the African American and Hispanic populations, the proportion of all births that are African American or Hispanic are projected to increase sharply in the next decade. Given that the rates for early prenatal care are lower and the incidence of low birth weight is higher in these populations, the subsequent demand for related health services can be expected to rise.

### *Compositional Attributes: Sociocultural Characteristics*

A number of compositional variables have implications for fertility patterns. Fertility differences by the educational attainment of the mother are even greater than those for race and ethnicity. Women in the 35 to 44 age range with less than a high school education will have the most children over their lifetimes, followed by high school graduates and college graduates, with those with an advanced degree recording half the number of births over their lifetimes as those without a high school diploma. College-educated women in particular have exhibited a dramatic shift to a later age of childbearing over the past 35 years.

Fertility levels vary widely by labor force status and income level. Women in the labor force have a GFR of 52.8 versus 90.0 for those not in the labor force. In regard to income, those at the lowest income, less than $10,000, have a GRR of 87.2 compared to that of 59.3 for women in the highest income range ($75,000 and more).

It was noted previously that Americans are having fewer children and having them later in life. Many couples, in fact, are opting to not have children at all. Lifestyles are thought to play a role in this phenomenon. More women are in the workforce and many have demanding careers. Further, the cost of raising children is a deterrent to many. Even groups noted for the relatively high birth rates in the past (e.g., African Americans) are exhibiting lower reproductive rates. Box 7.4 illustrates the trend in annual births for the US.

One lifestyle-related factor is changing attitudes toward children. It has been several decades since many parents considered their children as "insurance" for the future and planned to depend on them in old age. Similarly, it has been a long time since many Americans in agricultural communities felt the need to procreate in order to produce extra hands to work the farm. Today's attitudes with regard to having children reflect a rethinking of the importance of reproducing.

Another consideration in the United States is the impact of social psychological factors on fertility. This is another way of writing that reproduction is "trendier" at some points in history than others. The predisposition of couples to have children is influenced by societal expectations, current fashion, and competing values. Children in the United States were viewed as having a different function at the beginning of the twentieth century than they have now. In any society, the perceived function of children influences the value placed on progeny, which in turn influences the propensity to reproduce. In some societies, such a high value is placed on children (or on the connotations of reproduction for masculinity and femininity) that other factors such as changing socioeconomic status may be overridden.

***Variations in fertility rates over time can be attributed to a number of factors, including social, economic, and political factors.***

### *Social Conditions*

Variations in fertility rates over time can be attributed to a number of factors. These include social, economic, and political factors, not to mention more practical matters such as access to contraceptives. In the United States and other postindustrial countries, increases in educational levels and standards of living have contributed to declining fertility since the beginning of the twentieth century. This trend also was influenced by a major shift in the economic structure during this time period. Agrarian economies place a premium on large families; in industrial economies, large families become a liability. In addition, growing numbers of women have entered the labor force, particularly since World War II, further discouraging and/or preventing the establishment of large families.

Within this general trend, fluctuations can be identified that reflect social, economic and political conditions. Birth rates have historically declined during periods of economic uncertainty (e.g., the Great Depression) and increased during periods of prosperity (e.g., the post-WWII boom). Similarly, birth rates are typically lower during wartime than peacetime. Although government policy in the United States typically has not been utilized as a mechanism for controlling fertility, it has been argued that the federal tax structure historically has favored families with children. In some other societies (e.g., Prussian Germany and the People's Republic of China), fertility levels have been directly controlled by the ruling government.

**Box 7.4: Annual Births, United States 1945–2020**

| *Year* | *Births (in millions)* |
|---|---|
| 2020 | 3.6 |
| 2015 | 3.9 |
| 2010 | 4.0 |
| 2005 | 4.1 |
| 2000 | 4.0 |
| 1995 | 3.9 |
| 1990 | 4.1 |
| 1985 | 3.7 |
| 1980 | 3.6 |
| 1975 | 3.1 |
| 1970 | 3.7 |
| 1965 | 3.8 |
| 1960 | 4.3 |
| 1955 | 4.1 |
| 1950 | 3.6 |
| 1945 | 2.9 |

*Source* National Center for Health Statistics

Box 7.5 presents the number of births and the fertility rate for the time period 1910–2020. As can be seen there is a great deal of elasticity in both phenomena. Much of the variation over time can be correlated with social, economic and/or political events. Examples presented in the graph include periods of recession and the 1918 influenza epidemic. No doubt other events could be discerned by reading between the lines. It is important to note that, since the mid-1960s the number of births has steadily increased, yet the fertility rate has continued to decline. Although more babies were being born, the population base had increased so much as to reduce the rate.

**Box 7.5: Annual Births and Fertility Rates United States 1910–2020**

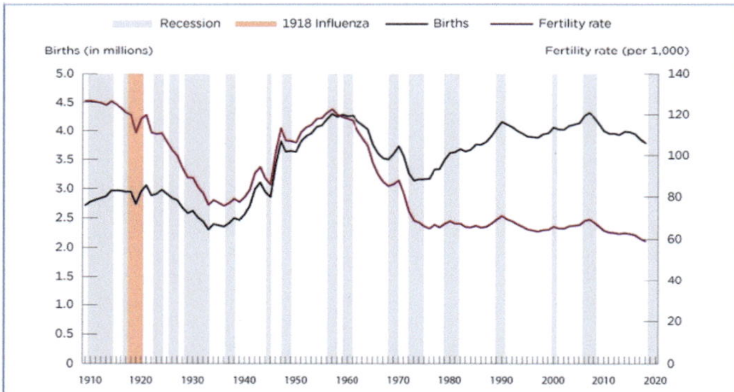


Note: The fertility rate is measured as births per 1,000 females ages 15 to 44.
Source: National Center for Health Statistics; U.S. Census Bureau, Population Estimates; and National Bureau of Economic Research.

The pattern illustrated in Box 7.6 is also reflected in the total fertility rate depicted in Box 7.7. Until around 1965 the US population exhibited a total fertility rate above the replacement level. However, that rate has remained below the replacement level ever since. Generations X, Y and Z all exhibit rates below replacement, Gen Z flirting with that rate in recent years.

**Box 7.6: Total Fertility Rate United States: 1910–2021**

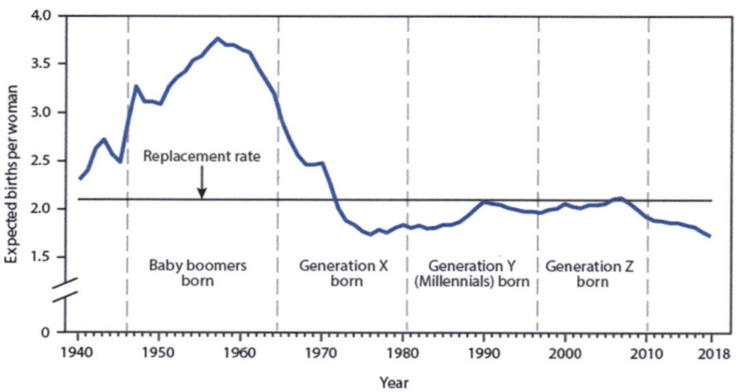


*Source* Centers for Disease Control and Prevention

The above discussion does not present an exhaustive list of the factors having potential impact on fertility, and the factors worthy of concern vary by service or market area. For example, other racial and ethnic group data would be of interest in the Southwest or on the West Coast. Compositional change (e.g., changing age or racial/ethnic composition) over time is equally important given the population redistribution patterns underway in the United States. Substantial increases in the number of births are possible over relatively short periods of time (three to five years) in rapidly growing areas. From an applied perspective, the importance of the above information lies in its salience for the level of need for birth-related services.

Box 7.7 presents age-specific fertility rates for four years: 1960, 1980, 2000 and 2020. The variation in age-specific birth rates is marked. During the post war baby boom all rates were high, with peak figures at ages 20–24 and 25–29. Twenty years later, during the baby bust, almost all rates had fallen, and by a large margin. However, the figures for 2000 show a shift in pattern. Rates for ages 20 and above all show an increase from those seen in 1980 and for the ages 30 and above the upward movement is substantial. The TFR rose between 1980 and 2000 in part due to older "baby boom" women belatedly bearing children.

**Box 7.7: Age-Specific Birth Rates (per 1000 females 15–44) United States 1960–2020**

| *Age category* | *1960 Rate* | *1980 Rate* | *2000 Rate* | *2020 Rate* |
|---|---|---|---|---|
| 10–14 | 0.8 | 1.1 | 0.9 | 0.2 |
| 15–19 | 89.1 | 53.0 | 47.7 | 15.3 |
| 20–24 | 258.1 | 115.1 | 109.7 | 62.8 |
| 25–29 | 197.4 | 112.9 | 113.5 | 90.0 |
| 30–34 | 112.7 | 61.9 | 91.2 | 94.8 |
| 35–39 | 56.2 | 19.8 | 39.7 | 51.7 |
| 40–44 | 15.5 | 3.9 | 11.8 | 11.8 |
| 45–49 | 0.9 | 0.2 | 0.5 | 0.9 |

*Source* National Center for Health Statistics

A number of important fertility-related trends can be identified for the United States population. As can be seen, the number of births increased from 2.9 million in 1945 to 4.3 million annually for 1957 through 1961. The period from 1946 to 1964 is generally recognized as the era of the post-World War II baby boom; until 1989, 1964 was the last year in which there were at least 4.0 million births in the United States. The interval 1965–1972 is seen as the transition to the *baby bust* that lasted from 1972 to 1978. Although some regard post-1978 fertility as evidence of a baby boomlet, a better explanation is to note that there was an *echo baby boom*. In other words, there was an increase in births due to the rapid rise in the number of potential and then actual mothers as the early baby boomers reached their childbearing years.

***Birth rates have steadily declined since 1960 for women of all ages, although women 35 and older have exhibited an upward trend in fertility since 2000.***

After 1987 there was an increase in the number of births that could not be explained by the echo effect. The number gradually increased through the end of the twentieth century and continued to rise in the early part of the twenty-first century. By 2007, the figure for annual births had reached 4.3 million, a number not realized since the 1957–1961 period. Despite this increase in number of births (the numerator), the fertility rate declined because the population (the denominator) was so much larger. In any case, this increase in births was short lived, with the figure for 2020 dropping to 3.6 million, a figure not seen since the 1970s.

As noted earlier, a TFR of 2.1 is considered to be replacement level fertility. Very low TFRs, less than 2.1, over a longer period of time results in an aging population and eventual population decline if there is not a counter-balancing immigration flow in place. The TFR for the US has been below 2.1 since the mid-1970s, and without significant immigration the US population would achieve zero population growth and subsequent population decline. China is a "poster child" for the consequences of decades of below replacement fertility and no immigration. Its population is aging rapidly and will begin to decline in size in the next 20 years.

There are several trends in factors related to birth outcomes that are important to consider. These factors help explain the hows and whys of fertility variations and trends. With regard to the likelihood of conception, several trends are worthy of note. Not only has contraceptive use increased since the 1970s, but the pattern of use has changed over time. Nearly 62% of women of childbearing age use some type of nonsurgical contraceptive (e.g., oral contraception, intrauterine device [IUD] or diaphragm). Nearly 17% have been sterilized (US Bureau of the Census 2010, table 97). Reliance on the pill, IUD, and diaphragm has declined since the 1970s, while sterilization as a means of contraception has become increasingly common. There remains a large number of women who do not use contraceptives, leaving them exposed to wanted and unwanted pregnancies.

A change in the average age at first intercourse can have important implications for the health of a population. For example, premarital intercourse on the part of teenagers has increased markedly since the 1970s. Earlier age of first intercourse results in a rise in the risk of pregnancy and increases exposure to sexually transmitted diseases. Despite the increasing younger age of first intercourse, the birth rate for women 15–19 has steadily declined largely due to increases in the proportion using contraceptives.

***An unprecedented number of births take place out of wedlock in the US today, with two-thirds of African American mothers being unmarried.***

Another major trend in fertility variation relates to children born out of wedlock. In 2008, less than 41% of all births in the United States were to unmarried mothers. Approximately 72% of all African American births were to women who were unmarried, but only 28% of white births were so classified (Martin et al., 2010). While the proportion of African American births occurring out of wedlock appears to have leveled off (albeit at a very high rate), the proportion of white and Hispanic births to unmarried mothers continues to increase. When the age of the mother is accounted for the race and ethnic differential narrows. At ages 15–19, 82% and 98% of all white and African American births, respectively, are to mothers who are unmarried. At ages 30–34, 1% of white births and 19% of African American births are to unmarried mothers. Box 7.8 presents recent data on the percentage of all births to mothers who are unmarried.

**Box 7.8: Total Births and Births to Unmarried Mothers United States 1960–2020**

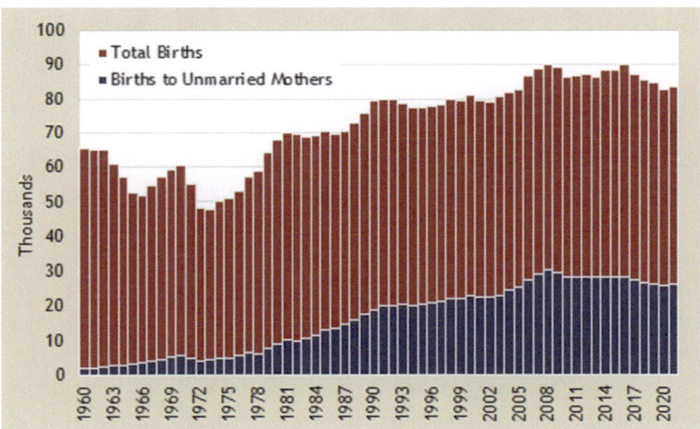


*Source* National Center for Health Statistics

One of the more controversial trends related to fertility in the United States is the growth in the number of annual abortions since that procedure was declared legal in the 1970s. As described in Box 11.5 in Chap. 11, the number of legal abortions climbed steadily with the liberalization of abortion laws from 200,000 in 1970 to a high of around 1.6 million in 1980. Since 1980 the annual number has steadily declined, representing a decrease of about 400,000 annually from the early 1980s. Services related to the performance of induced abortions involve a variety of healthcare components, including medical, counseling and contraceptive services. The changing number of abortions performed is discussed later in the chapter on US demographic trends.

***The number of legal abortions performed annually increased dramatically after 1970 but has been declining since the 1980s.***

The changes that are taking place in US society have significant implications for fertility. At the same time, changes in fertility patterns have important implications for US society (the two-way street). With regard to the latter, the decline in fertility for the short term portends an aging population, as there are fewer births to shore up the younger age cohorts. At the same time the US population will become more female heavy as it ages. Differentials in fertility will also be to the detriment of the non-Hispanic white population. Declines in white fertility juxtaposed with high non-white fertility will alter the racial/ethnic composition of the population. The impact of declining fertility on education has previously been noted, with the lack of college-aged youth representing a crisis for higher education. Lower fertility also creates challenges for the economy. In an environment where labor shortages continue to be reported in the wake of the COVID-19 pandemic, the dearth of youth for the foreseeable future bodes ill for an economic system that depends on a ready source of entry-level workers.

## Data Sources on Fertility

Fertility data are drawn from a variety of sources, although official vital statistics registries represent the most reliable source of information on fertility. Relatively high-quality birth registration systems exist in each state, and the standard birth certificate includes a variety of data on the characteristics of the child, mother, and father. Data from these state systems are compiled at the national level by the National Center for Health Statistics. (Refer to Box 6.2 for the elements found in the standard birth certificate.)

The conditions of the newborns (e.g., birth weight and Apgar scores) can be determined along with compositional traits (e.g., marital status and age) of the mothers. This information is typically gathered from birth certificates. For statistical purposes, births are assigned to both the mother's *place of residence* and *place of occurrence*. For most purposes, the place of residence is the most important consideration. Fertility rates and associated statistics are relevant for the community in which the mother resides and not for the place where the birth occurred, especially if the delivery takes place outside the community of residence.

***The standard birth certificate is the primary source of data on births and the characteristics of mothers and their babies.***

In addition to the birth registration system, sample surveys are a source of data on fertility-related behavior. Surveys provide information on such issues as contraceptive use, infertility, and breast-feeding practices. Surveys like the

National Survey of Family Growth conducted by the National Center for Health Statistics make it possible to track trends in fertility-related behavior.

The US Census Bureau collects data on a limited number of fertility issues through both the American Community Survey (ACS) and the Current Population Survey (CPS). In the ACS, women are asked whether or not they had given birth within the previous year. The CPS has a more extensive list of fertility inquiries, including questions with regard to actual and expected fertility are asked and cross-tabulated by other demographic factors. See Chapter 7 for more details.

## Key Points

- Fertility is a social process requiring the biological interaction of two persons within a defined economic, social and/or political context.
- Fertility is measured in terms of the number of births that occur within a population and the fertility rate it represents.
- The crude birth rate is the most basic measure of fertility and is calculated by dividing the total number of births for a given year by the midyear population.
- The general fertility rate represents a refinement of the crude birth rate, since it uses the population at risk for pregnancy as the denominator.
- The total fertility rate is sometimes utilized as a summary measure for age-specific fertility rates.
- Replacement-level fertility, the number of births required for females to exactly replace themselves, is 2.1 births per woman over a lifetime.
- Among the issues related to reproductive health outcomes are premature births and low-birth weight deliveries.
- A well-established model of fertility levels attributes the fertility rate to exposure to intercourse, exposure to contraception, pregnancy outcomes.
- Variations in fertility rates over time can be attributed to a number of factors, including social, economic, and political considerations.
- Birth rates have steadily declined since 1960 for all age groups, although women 35 and older have exhibited an upward trend in fertility since 2000.
- The number of legal abortions performed annually increased dramatically after 1970 but has been declining since the 1980s.
- An unprecedented number of births take place out of wedlock in the US today, with two-thirds of African American mothers being unmarried.
- The standard birth certificate is the primary source of data on births and the characteristics of mothers and their babies.

## Discussion Questions

7.1 African Americans continue to have worse pregnancy outcomes than most other racial and ethnic groups. This situation is often attributed to the lower socioeconomic status of black women. However, when socioeconomic status is controlled for, African American women still record higher rates of infant mortality and maternal mortality. What do you think accounts for the fact that African American women continue to report pregnancy outcomes that are comparable to those in undeveloped countries?

7.2 Americans report fertility rates that are well below replacement levels. The figure would be even worse, were it not for the high birth rates reported by certain racial and ethnic groups. What are the implications of our population not reproducing itself? Are there actions that can be taken to reverse this trend?

## Exercises

*Exercise 7.1: Creating a County Fertility Profile*

For this assignment students should select a county and gather information on the fertility characteristics of that county. At a minimum, the following factors should be considered:

- Current birth and fertility rates
  - Crude birth rate
  - General fertility rate
  - Total fertility rate
- Historical trends related to each rate above.

Students should present their conclusions concerning fertility patterns/trends.

*Exercise 7.2: Analyzing Trends in Age-Specific Birth Rates*

Students should analyze the data in Box 7.8 detailing age-specific births for 1960–2020 to identify patterns of fertility characterizing different age groups in US society. Students should use information on the factors that determine fertility rates to analyze the patterns displayed in the table.

*Exercise 7.3: Mining Birth Certificate Data**

As an applied demographer you have been asked to address the high infant mortality rate displayed by county residents. You have been given the information from the birth certificates filed with the county for the past five years, and you should use this as the basis for analyzing the situation and generating ideas for reducing infant mortality in that community.

With your group members, examine the birth certificates and then:

(1) Identify the data elements that might be useful in understanding the incidence of infant mortality among this population
(2) Speculate on how the characteristics exhibited by the affected community might contribute to high infant mortality rates
(3) Suggest ways in which this information might be used by public health officials, government agencies and other parties to address the problem of high infant mortality

State any assumptions you are making and be able to explain your responses in demographic terms.

*See Box 6.2 for a list of information on the standard birth certificate.

## References

Davis, K., & Blake, J. (1956). Social structure and fertility: An analytic framework. *Economic Development and Cultural Change, 4*(3), 211–235.

National Center for Health Statistics. (2022). *Health United States, 2022*. National Center for Health Statistics.

US Census Bureau. (2010). *Statistical abstract of the United States*. US Government Printing Office.

## Additional Resources

Pressat, R. (2024). *Demographic analysis: Projections on natality, fertility and replacement*. Routledge.

Shoen, R. (2020). *Analyzing contemporary fertility*. Springer.

Yusuf, F., Martins, J., & Swanson, D. (2014). *Methods of demographic analysis*. Springer.

CHAPTER 8

# Demographic Processes: Mortality

**Abstract** This chapter defines and describes the mortality process, one of the three basic demographic processes. The importance of mortality to demographers and to society is reviewed. The various ways of classifying deaths and measuring mortality are described and an introduction to life tables provided. The factors that contribute to mortality are discussed, along with the implications of the ever-evolving causes of death in US society.

## Introduction

*Mortality* technically refers to the condition of death characterizing a population. Mortality analysis investigates the who, how, why, and when issues related to dying. Demographers have contributed greatly to the understanding of mortality both in terms of the development of mortality measures and the identification of patterns of mortality within the population. Mortality analysis also provides insights into the morbidity profile of the population and offers a window into the social, lifestyle and environmental contributors to health status and mortality.

The study of mortality has historically been important to population scientists for several reasons. As one of the three primary demographic processes its analysis is considered critical to an understanding of population growth and change. Mortality has served as a limiting factor for population growth, and patterns of mortality contribute to changing population composition. Thus, the current preponderance of seniors (and, conversely, the relative dearth of youth) in the US reflects the population's decreased mortality and increased longevity.



R. K. Thomas, *Demography: An Introduction to Population Studies*, Springer Texts in Social Sciences, https://doi.org/10.1007/978-3-031-56623-3_8

***As one of the three primary demographic processes an analysis of mortality is considered critical to an understanding of population growth and change.***

Mortality analysis is important because of what it could tell us about morbidity patterns within the population. Historically, there was a fairly close correlation between common maladies and common causes of death. The immediate cause of death was typically the primary cause of death, with few complicating factors involved. That connection can still be made today to a certain extent, in that the leading causes of death (heart disease and cancer) reflect common maladies within the population.

Contemporary population scientists place less emphasis on mortality analysis than they have in the past. As a component of population change, mortality has become less important than fertility and both have become less important than migration. Because of the preponderance of chronic disease within the US population, death certificates are less and less likely to capture the underlying disease. The situation is much more complicated than in the past, and analysts require a better understanding of the interaction of fertility, mortality and migration.

## Concepts and Measures

### *Concepts*

*Death* is defined as the complete cessation of life after a live birth has taken place. Deaths that occur prior to a live birth–fetal deaths–are allocated to a separate category of mortality study. Though the words "cessation of life" represents a simple, straightforward definition, medical advances are making the interpretation of death more complex. As the ability to artificially prolong life has improved, the distinction between life and death has blurred. The definition of death is constantly being rewritten in the light of medical and technological advances.

***Mortality refers to the level of death within a population as measured by the number of deaths and subsequent death rates.***

*Mortality* refers to the level of death within a population as measured by the number of deaths and death rates characterizing that population. In the US. extremely accurate data are available for mortality, and relatively timely information on deaths and death rates is available. Death rates are generally provided for the total population although it is not uncommon to present death rates for subsets of the population or to adjust death rates for standardization purposes (see below). Mortality data are generally reported based on the place of residence of the deceased, although data are also collected on the place of occurrence of this vital event.

*Infant mortality* refers to the number and rate of deaths recorded for infants under one year of age. The rate of infant mortality is an effective measure of not only health conditions but social, economic and environmental conditions within a population, thereby providing significant insights beyond just the occurrence of infant deaths. The infant mortality rate (IMR) is used to compare the health and well-being of populations across and within different geographic areas.

***Infant mortality refers to the number and rate of deaths recorded for infants under one year of age.***

The IMR can be divided into *neonatal* and *post-neonatal* components, with the former referring to deaths during the first 28 days of life and the latter reflecting deaths occurring from 29 days to one year. Deaths occurring prior to delivery are considered *fetal* deaths and are commonly referred to as miscarriages.

*Maternal mortality* refers to deaths experienced by mothers as a result of the birthing process. Maternal mortality was a major killer of women until the twentieth century when more sanitary obstetrical procedures were introduced. Both infant mortality and maternal mortality are further discussed later in the chapter.

The *causes of death* affecting a population are a major factor in determining the level of mortality. Populations in different times and places are subject to different causes of death. Knowing the number of people who died is one thing, but knowing what they died from provides valuable insights into the overall health status of the population and the types of health conditions that afflict it.

Information on the cause of death in the US is compiled from certificates filed with health authorities on the occasion of any death. Since virtually every death is accompanied by a death certificate, the information on cause of death is fairly complete.

***Maternal mortality refers to deaths experienced by mothers to the birthing process.***

There are, however, a number of issues related to the specification of cause of death. Given today's morbidity patterns, it is increasingly difficult to specify the ultimate cause of death. With a preponderance of chronic diseases, it is often the case that death can and should be attributed to a factor other than the underlying cause of death. Patients with AIDS do not typically die as a direct result of AIDS but due to system failure attributed to AIDS. Similarly, individuals affected by diabetes are often said to die from "complications of diabetes".

The inclusion of contributing factors to mortality has become increasingly useful as chronic disease has come to be more common than acute disease. While the immediate cause of death in a particular case may be kidney failure, it is useful to know that diabetes was the underlying cause. Similarly, obesity, while not an immediate cause of death, is increasingly being listed as a contributing factor. We have actually developed a better appreciation of the role of obesity in a variety of health problems through the analysis of death certificates.

While death certificates represent a significant source of data for mortality analysis, there are issues that require caution in their use. There is not universal agreement as to the determination of which factor is the immediate cause of death. There are, in fact, differences that exist from community to community with regard to the classification of contributing and proximate factors.

### *Measures of Mortality*

Counts. The most basic way to measure mortality is simply to count the number of deaths. Such counts are usually based on a one-year period and may be reported for the nation as a whole, states, metropolitan areas or smaller geographic areas. Compiling death counts over a period of years has helped identify trends with regard to increases or decreases in mortality. Deaths are also cross-classified by the medical, social, and economic characteristics of the deceased (e.g., cause of death and age at death).

***The most basic way to measure mortality is simply to count the number of deaths.***

Using a simple count of deaths in the analysis of mortality has several shortcomings. As in the case of fertility analysis, the comparison of deaths among geographic areas or over time generally is not very useful, given the various sizes of the populations reporting these deaths. It is usually of little value to compare the number of deaths in Community X (population 10,000) with the number of deaths in Community Y (population 100,000). Because of the need to compare the level of mortality for different populations and over time, demographers have developed a number of rates for this purpose. Box 8.1 presents common measures used for mortality analysis.

Death Rates.[1] The simplest measure used is the *crude death rate* (CDR). Like the crude birth rate discussed in Chap. 7, this rate expresses mortality as the number of deaths per 1000 population during a particular year (e.g., 6.5 deaths per 1000 people). The number of deaths, particularly for small populations, may fluctuate widely from year to year, thus researchers may wish to calculate a 3-year average for the number of deaths.

[1] "Death rate" is used as a generic term for any measurement of mortality. "Mortality rate" is a more formal term that typically involves a calculation based on a specific formula.

***The crude death rate expresses mortality as the number of deaths per 1000 population.***

Just as everyone in a population is not at risk of pregnancy, not everyone in a population is at equal risk of death. Lumping all deaths into one crude death rate limits the measure's usefulness. In order to refine this measure, age-specific death rates (ASDRs) can be generated. By determining the death rate for each age cohort, it becomes possible to determine which portions of the age distribution account for the observed mortality. ASDRs are usually calculated for 5-year age intervals, though 3- or even 1-year intervals may be used for more detailed analyses.

The infant mortality rate (IMR) represents a special case of age-specific death rates. The IMR is expressed as a 1-year rate and is examined separately because of the greatly increased probability of dying during the first year of life compared to subsequent ages. Persons under age 1, for example, are 20 times more likely to die in a given year than someone in the 1–4 year-old cohort.

Infant mortality, like mortality in general, is not evenly distributed throughout the population but affects some groups more than others. African Americans and American Indians exhibit inordinately high infant mortality rates, while Asian Americans and Hispanics report lower that average rates.

Neonatal deaths refer to infant deaths during the first 28 days of life. Four conditions—congenital abnormalities (birth defects), low birth weight, respiratory distress, and maternal complications—account for over half of all infant deaths. Post-neonatal deaths are those that occur during from day 29 to one year of age. Sudden infant death (SIDS) is a major factor in post-neonatal deaths, accounting for some 30% of deaths during this period.

**Box 8.1: The Calculation of Mortality Rates**

The calculation of mortality rates is relatively straightforward and the requisite data are usually readily available. Death data are available from vital statistics registries, and population figures (rate denominators) can be drawn from Census Bureau data, state demographic centers or commercial data vendors.

$$\text{Crude death rate(CDR)} = \frac{\text{Number of Deaths in year } X}{\text{Midyear population in year } X} \times 100$$

To avoid anomalies in the data related to a single year (e.g., results of an epidemic or natural disaster), it is common to utilize a three-year average with the population for the midpoint of the middle year used for the denominator.

$$\text{Age-specific death rate(ASDR)} = \frac{\text{Deaths to persons age } X \text{ to age } X + 5 \text{ in year } Y}{\text{Number of persons age } X \text{ to age } X + 5 \text{ in year } Y} \times 100$$

with $X$ to $X + 5$ signifying the five-year interval. Three-year averages for deaths may be used as numerators if there is substantial year-to-year variation in the number of deaths.

$$\text{Infant mortality rate(IMR)} = \frac{\text{Number of deaths to persons under 1 year of age in year } X}{\text{Live births in year } X} \times 1000$$

Since infant deaths are a relatively rare occurrence, three-year averages for infant deaths may be used, especially since there is likely to be significant year-to-year variation. The same method may be applied to neonatal and post-neonatal death calculations.

$$\text{Neonatal mortality rate} = \frac{\text{Number of deaths to persons 1 to 28 days of age in year } X}{\text{Live births in year } X} \times 1000$$

$$\text{Post-neonatal mortality rate} = \frac{\text{Number of deaths to persons 29 days to year in year } X}{\text{Live births in year } X} \times 1000$$

$$\text{Cause-specific death rate (CSDR)} = \frac{\text{Number of deaths from a specific cause in year } X}{\text{Midyear population in year } X} \times 1000$$

$$\text{Maternal mortality rate (MMR)} = \frac{\text{Number of women dying during childbirth in year } X}{\text{Live births in year } X} \times 1000$$

As can be seen from Box 8.2, in the US today the likelihood of dying is different for each age cohort. The risk of death is relatively high during the first year of life but drops dramatically for children and teenagers. The likelihood of death increases slowly for each age cohort after the teen years with death rates beginning to accelerate when the population reaches “middle age”. Not surprisingly, the greatest risk of death is exhibited by the oldest age cohorts.

This trend is somewhat different from that 20 years ago. The infant mortality rate is less pronounced, and the elderly death rate is greatly augmented today. This indicates that death in contemporary US society has become increasingly concentrated within the oldest age groups.

**Box 8.2: Age-Specific Mortality Rates United States 2020**

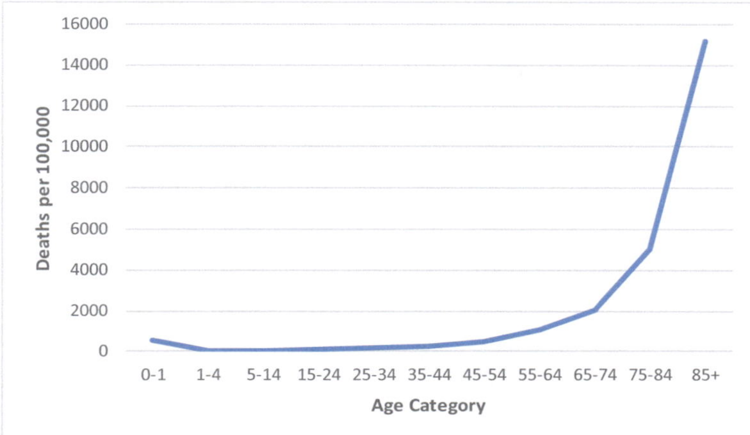


*Source* National Center for Health Statistics

Significant differences exist between males and females with regard to mortality. Males exhibit higher mortality rates for virtually all causes and at all ages. As indicated in Box 8.3, the all-cause death rate for males is about one-third higher than it is for females (864 per 100,000 compared to 616). Similar disparities can be seen for each of the leading causes of death. The one exception is deaths from Alzheimer's disease. This, of course, is due to the fact that women generally live longer men, putting them at greater risk for Alzheimer's.

**Box 8.3: Sex-Specific Death Rates United States 2020**

| | *Death rate (per 100,000)* | |
|---|---|---|
| *Cause of Death* | *Males* | *Females* |
| Heart disease | 209 | 130 |
| Cancer | 181 | 130 |
| Accidents | 68 | 32 |
| Chronic lower respiratory diseases | 45 | 38 |
| Stroke | 38 | 37 |
| Alzheimer's disease | 25 | 35 |
| Diabetes | 26 | 17 |
| Total | 864 | 619 |

*Source* National Center for Health Statistics

## Cause of Death

While the age at death provides useful information, a more complete picture of mortality patterns can be generated by examining cause-specific deaths. The method for calculating cause-specific death rates (CSDR) is similar to that for age-specific death rate, with the numerator being the number of deaths attributed to a particular cause.

The death rate for various causes provides information on the types of health conditions faced by that population. In addition, these data indicate a great deal about the conditions of life within that particular group and about the demographic attributes of a population. For example, a high rate of acute conditions relative to chronic conditions would suggest that the population was relatively young or existed in primitive conditions. On the other hand, a preponderance of chronic conditions would suggest an older population and/or exposure to circumstances that resulted in chronic disease.

At the same time, knowledge of the demographic attributes of a population provides clues to the health conditions that could be expected. A relatively young population would generate more acute health conditions while an older population would be expected to exhibit more chronic health conditions. Further, knowing the sex ratio or the racial composition of the population provides further clues to the diseases responsible for deaths in that population. Box 8.4 presents the leading causes of death for the US in 2020.

**Box 8.4: Leading Causes of Death United States 2020**

| *Cause of death* | *Number* | *Percent* |
|---|---|---|
| Heart disease | 690,892 | 20.5 |
| Cancer | 598,932 | 17.8 |
| COVID-19 | 345,323 | 10.3 |
| Accidents | 133,382 | 4.0 |
| Stroke | 159,050 | 4.7 |
| Chronic lower respiratory diseases | 151,631 | 4.5 |
| Alzheimer's disease | 133,832 | 4.0 |
| Diabetes | 101,106 | 3.0 |
| Pneumonia/influenza | 53,495 | 1.6 |
| Kidney disease | 52,260 | 1.6 |
| Suicide | 44,834 | 1.3 |
| Total | 3,358,814 | |

*Source* National Center for Health Statistics

### *Potential Years of Life Lost*

Years of Potential Life Lost (YPLL) is a widely used measure of the rate and distribution of premature mortality. Measuring premature mortality, rather than overall mortality, focuses attention on deaths that might have been prevented. YPLL emphasizes deaths of younger persons, whereas statistics that include all mortality are dominated by deaths of the elderly. For example, using YPLL-75, a death at age 55 counts twice as much as a death at age 65, and a death at age 35 counts eight times as much as a death at age 70.

The excess mortality rate among Americans ages 15–64 was 58 per 100,000 people in this age group in 2020—more than double that of the next closest peer nation, the United Kingdom (25 per 100,000). Nearly half (48%) of excess deaths in the US were among people younger than 75, compared to 18% for Belgium, a country with a comparable overall excess mortality rate.

Excess potential years of life lost (up to age 75) is a measure of excess mortality and is used to compare differences in disease burden and longevity across countries. The analysis finds that the US had 1171 excess potential years of life lost up to age 75 per 100,000 people ages 0–74, which is over twice the rate of premature excess mortality in the next closest country (488 per 100,000 people). This approach, which follows OECD methods, may understate premature excess mortality in 2020, as excess deaths over age of 75 in 2020 were also premature compared to a typical year. Prior to 2020, the US already had the highest rate of premature deaths among peer countries. This analysis shows the gap in premature mortality rates between the US and peer countries has increased due to the COVID-19 pandemic.

The 3.6 million potential years of life in 2020 in the US was driven in part by racial disparities. American Indian and Alaska native, African American, Native Hawaiian and other Pacific Islander, and Hispanics report over three times the premature excess death rate. Thirty percent of the total excess potential years of life lost in the US are among African Americans, and 31% are among Hispanics.

## Data Standardization

For many measures of mortality the rates may be misleading if the population under study is not a "standard" population. When comparing two communities in terms of their mortality profiles, the various rates discussed above are useful if the population composition is similar. However, in many cases, distinctly different populations may affect the rates and result in misleading comparisons. For example, the crude death rate for Florida in 2020 was 12.0 per 1000 population, a figure much higher than the CDR of 10.5 per 1000 for Nevada in that same year. This would suggest that Nevada is a healthier state than Florida, even though common sense might suggest that this is not the case. Indeed, when the death rate is standardized to take into account differences in age structure, the age-adjusted death rate for Florida is 7.4 and for Nevada is 8.5.

## Life Tables

Although inappropriately named, life tables provide demographers, health planners,, insurance companies and others involved in risk management with a great deal of useful information. Life tables are a mechanism for combining death data and rates into a summary measure of mortality. Age-specific mortality rates are transformed and combined in such a way as to generate life expectancy rates.

*Life expectancy*, in general terms, is the average number of years a hypothetical group or cohort of persons born today or alive at a particular point in time could be expected to live if current age specific death rates remain constant throughout their lifetimes. In other words, as the hypothetical cohort ages it is subject to probabilities of dying specific to that age interval, which in turn are based upon current ASDRs. Life expectancy projections are made by adjusting ASDRs to match the assumed levels for some time in the future. Box 8.5 presents an abridged life table for the United States.

Life tables provide a great deal of information about mortality conditions, and it is possible to trace trends in mortality by assembling life table data for more than one time period. Comparative life table analysis can yield information on life expectancy differentials by sex and race for any age. For example, though life expectancy at birth is more than five years greater for females than it is for males, the difference narrows to four years at age 40, three years at age 60, and two years at age 80. When one looks jointly at race and sex, larger differentials emerge.

***Life tables provide a great deal of information about mortality conditions and the mortality experience of a population.***

Not only do these data have serious social, economic, and political implications, but the differentials indicate that two very different populations exist vis-à-vis the healthcare system. Healthcare providers and policy makers alike must take differential life expectancies into consideration in performing their respective tasks. Thus, using the black male versus white female example, out of 100,000 white females born over 85,000 will be alive at age 65. For black males, only 57,000 will be alive, or only about two thirds of the white female population.

Based on the above data and other observations, it can be said that the African American male population is considerably younger than the remaining three sex/race groupings, African American female, white male, and white female. Coupled with additional psychographic and behavioral information, specific strategies regarding service development and promotion could be formulated for this very different subpopulation.

A final use for life tables involves the calculation of survival ratios. These ratios are a means of determining the proportion of persons surviving from one age interval to the next. Once calculated, they can be used to determine what proportion of persons alive now can be expected to be alive at some point in the future given the assumption that death rates remain relatively constant.

The survival ratio for persons 75 years of age living to 80 is 0.716. That is, about 72% of the persons who were alive at age 75 will also be alive at age 80. Cause-specific life tables further allow the measurement of the effect of the hypothetical removal of certain causes of death on overall life expectancy. It is not unusual to find life table calculations assuming that heart disease or cancer is eliminated as a cause of death. Clinical researchers and healthcare administrators alike have a vested interest in the survival ratios of patients on whom various procedures are performed.

While most life tables are constructed at the national or state levels, methods have been developed to generate substate life expectancies. Furthermore, simple life tables can be expanded to include joint probabilities of dying and having a certain condition, such as a disability. For example, life expectancy for those who were 70 in 1997 report life expectancy of 84 years for those without a disability and 82 years for those with a disability. Research along these same lines has resulted in the creation of a composite indicator, the *healthy life-year* (HeaLY). For this indicator, life expectancy and morbidity data are combined to produce figures for life-years lost due to a variety of conditions (Hyder et al., 1998).

**Box 8.5: Sample Abridged Life Table for the United States**

| *Age* | *Probability of dying between ages* x *to* x + n<br>$_nq_x$ | *Number surviving to age* x<br>$l_x$ | *Number dying between ages* x *to* x + n<br>$_nd_x$ | *Person-years lived between ages* x *to* x + n<br>$_nL_x$ | *Total number of person-years lived above age* x<br>$T_x$ | *Expectancy of life at age* x<br>$e_x$ |
|---|---|---|---|---|---|---|
| 0–1 | 0.006879 | 100,000 | 688 | 99,398 | 7,784,998 | 77.8 |
| 1–5 | 0.001174 | 99,312 | 117 | 396,970 | 7,685,600 | 77.4 |
| 5–10 | 0.000727 | 99,196 | 72 | 495,784 | 7,288,630 | 73.5 |
| 10–15 | 0.000898 | 99,124 | 89 | 495,452 | 6,792,846 | 68.5 |
| 15–20 | 0.003251 | 99,035 | 322 | 494,460 | 6,297,395 | 63.6 |
| 20–25 | 0.004869 | 98,713 | 481 | 492,387 | 5,802,935 | 58.8 |
| 25–30 | 0.004865 | 98,232 | 478 | 489,966 | 5,310,547 | 54.1 |
| 30–35 | 0.005551 | 97,754 | 543 | 487,457 | 4,820,581 | 49.3 |
| 35–40 | 0.007433 | 97,211 | 723 | 484,370 | 4,333,124 | 44.6 |
| 40–45 | 0.011588 | 96,489 | 1118 | 479,837 | 3,848,755 | 39.9 |
| 45–50 | 0.017540 | 95,371 | 1673 | 472,927 | 3,368,918 | 35.3 |
| 50–55 | 0.025802 | 93,698 | 2418 | 462,770 | 2,895,990 | 30.9 |

| | | | | | | |
|---|---|---|---|---|---|---|
| 55–60 | 0.036299 | 91,280 | 3313 | 448,575 | 2,433,221 | 26.7 |
| 60–65 | 0.055819 | 87,967 | 4910 | 428,282 | 1,984,646 | 22.6 |
| 65–70 | 0.082066 | 83,057 | 6816 | 399,173 | 1,556,364 | 18.7 |
| 70–75 | 0.125036 | 76,241 | 9533 | 358,595 | 1,157,191 | 15.2 |
| 75–80 | 0.188740 | 66,708 | 12,590 | 303,365 | 798,596 | 12.0 |
| 80–85 | 0.288884 | 54,117 | 15,634 | 232,350 | 495,231 | 9.2 |
| 85–90 | 0.420212 | 38,484 | 16,171 | 151,473 | 262,881 | 6.8 |
| 90–95 | 0.575974 | 22,312 | 12,851 | 77,357 | 111,408 | 5.0 |
| 95–100 | 0.733375 | 9461 | 6938 | 27,543 | 34,051 | 3.6 |
| 100 and over | 1.000000 | 2523 | 2523 | 6508 | 6508 | 2.6 |

$_nq_x$ = the proportion of persons alive at the beginning of the interval who died during the interval
$l_x$ = the number surviving from one age interval to the next after deaths are subtracted
$_nd_x$ = the number of deaths that occur to that age interval during the specified time period
$_nL_x$ = the number of person-years lived during that interval
$T_x$ = the reverse sum of $_nL_x$ values
$E_x$ = the life expectancy at the specified age
*Source* National Center for Health Statistics (2006). United States Life Tables, 2003. *National Vital Statistics Reports* Vol. 54, No. 14

## US Mortality Trends

Mortality is a dynamic process, influenced by a number of demographic factors and subject to significant fluctuations. For this reason there is a certain amount of "elasticity" in mortality rates. While the trend has generally been downward since the beginning of the twentieth century, short-term fluctuations tend to be common.

### *Deaths*

The total number of annual deaths in the United States, as indicated in Box 8.7, increased steadily during the twentieth century. However, by 2005 this trend had moderated and the number of annual deaths stabilized at around 2.42 million per year, or around one million more deaths annually compared with 1935. This number increased slightly between 2006 and 2010.

### *Death Rates*

More significant than the number of deaths, however, is the mortality rate since it is adjusted for population size. The crude death rate in 1900 was 17.2 per 1000. This figure declined dramatically during the first one-half of the twentieth century, with a rate of 10.8 recorded in 1940. The death rate continued to decline during the century, dropping to around 8 per 1000 in 2010. Since then, the crude death rate has unexpectedly increased and stood at 8.4 for 2020.

At the beginning of the twentieth century life expectancy was about 49 years, with women outliving men by about three years.[2] Over time, life expectancy increased dramatically to over 59 years in 1930, 68 years in 1970 and over 75 years in 1990. In 2006, life expectancy in the US was over 77 years and at that time projected to continue to rise (US Bureau of the Census 2010, table 102). During the same period of time the male/female differential in life expectancy widened, reaching seven years by 1970. The seven-year margin continued until 1990, followed by a gradual narrowing of the difference. Today, women outlive men by 5.0 years on the average. An interesting twist has arisen with regard to life expectancy in some geographic areas of the United States. Recent research has indicated that the trend toward greater longevity has been reversed for the populations in a number of US counties initially, with 4% of US men and 16% of US women experiencing *declining* life expectancy (Ezzati et al., 2008). This phenomenon has spread to the larger society to the point that the death rate for the US population overall has increased (after a century of decline).

In addition to realizing a substantial increase in life expectancy over the twentieth century and into the twenty-first century, the relative health condition at each age has improved. For example, recent research has allowed demographic researchers to extend life expectancy information to include healthy life years expected. For health planners as well as anyone else who wishes to better understand current demand for health services, particularly for the population age 50 and above, factoring in the health condition of the population that reaches age 50, 60 or 100 must take place in order to understand the market and how it might be changing. Box 8.6 describes the unexpected increase in US death rates.

[2] Demographers make a distinction between "life span" and "life expectancy". Life span refers to the maximum years that a species is able to live under ideal circumstances (i.e., over 100 years for humans). Life expectancy refers to the length of tine on the average that a human can expect to live.

**Box 8.6: Increasing Mortality in the US**

Epidemiologists have noted a marked increase in the all-cause mortality of middle-aged white non-Hispanic men and women in the United States since the late 1990s. This change reversed decades of progress in reducing mortality and was unique to the United States; no other developed country experienced a similar turnaround. The midlife mortality reversal was confined to non-Hispanics whites; blacks and Hispanics at midlife, and those aged 65 and above in every racial and ethnic group, continued to see mortality rates fall.

This increase for whites was largely accounted for by increasing death rates from drug and alcohol poisoning, suicide, and chronic liver diseases and cirrhosis. Although all education groups saw increases in mortality from suicide and poisoning and an overall increase in external cause mortality, those with less education saw the most marked increases.

Rising midlife mortality rates of non-Hispanic whites were paralleled by increases in midlife morbidity. Self-reported declines in health, mental health, and ability to conduct activities of daily living were reported. Increases in chronic pain and inability to work, as well as clinically measured deteriorations in liver function, all point to growing distress for the affected population.

**Box 8.7: Deaths and Death Rates United States 1970–2020**

| *Year* | *Deaths* | *Crude death rate* |
|---|---|---|
| 2020 | 3383* | 10.3 |
| 2015 | 2712 | 8.4 |
| 2010 | 2468 | 8.0 |
| 2005 | 2448 | 8.3 |
| 2000 | 2403 | 8.5 |
| 1995 | 2312 | 8.7 |
| 1990 | 2146 | 8.8 |
| 1985 | 2084 | 8.8 |
| 1980 | 1990 | 8.8 |
| 1975 | 1893 | 9.5 |
| 1970 | 1921 | 9.5 |

*It should be noted that the 2020 figure reflects the impact of the COVID-19 pandemic. Although death rates had started increasing prior to the pandemic. The 2020 figure includes an unknown number of "excess" deaths attributable to COVID-19

*Source* National Center for Health Statistics

**Box 8.8: Life Expectancy at Birth by Sex United States 1900–2020**

| *Year* | *Total* | *Males* | *Females* |
|---|---|---|---|
| 2020 | 77.3 | 74.2 | 79.9 |
| 2015 | 78.8 | 76.3 | 81.2 |
| 2010 | 78.7 | 76.2 | 81.0 |
| 2005 | 77.4 | 74.9 | 79.9 |
| 2000 | 76.8 | 74.1 | 79.3 |
| 1995 | 75.8 | 72.5 | 78.9 |
| 1990 | 75.4 | 71.8 | 78.8 |
| 1985 | 74.7 | 71.1 | 78.2 |
| 1980 | 73.7 | 70.0 | 77.5 |
| 1975 | 72.6 | 68.8 | 76.6 |
| 1970 | 70.8 | 67.1 | 74.8 |
| 1960 | 69.7 | 66.6 | 73.1 |
| 1950 | 68.2 | 65.6 | 71.1 |
| 1940 | 62.9 | 60.8 | 65.2 |
| 1930 | 59.7 | 58.1 | 61.6 |
| 1920 | 54.1 | 53.6 | 54.6 |
| 1910 | 50.0 | 48.4 | 51.8 |
| 1900 | 47.3 | 46.3 | 48.3 |

*Source* National Center for Health Statistics

One of the significant developments during the twentieth century in the US was the dramatic reduction in infant mortality rates. As can be seen in Box 8.9, early in that century infant mortality was a leading cause of death and contributed to a relatively short life expectancy. Between 1920 and 1940 the IMR was cut essentially in half and by 1970 it was halved again. Overall, a rate of 85 infant deaths per 1000 live births in 1920 was cut to 7 per 1000 by 2000. A similar but even more dramatic trend was recorded for maternal deaths, a leading cause of death for females at the beginning of the twentieth century. The maternal death rate of around 80 per 10,000 births was reduced to only 1 per 10,000 by the end of the twentieth century. The improvements in infant and maternal mortality rates can be attributed to a number of factors in addition to better medical care, including improved public health conditions and changes in demographic characteristics such as education and income. Moreover, the sharp reduction in the maternal mortality rate was directly responsible for the rapid increase in female life expectancy that took place in the middle part of the twentieth century.

Life expectancy is tied directly to mortality and is considered the converse of death rates. As the mortality rate goes down, years of life expectancy typically goes up. Conversely, if the mortality rate increases, we can expect a decline in life expectancy (as we are currently seeing in the US today).

The disappointing progress the US has made in reducing infant mortality is to a great extent a function of disparities in access to healthcare that exist between various groups. The relatively large infant mortality disparity between African Americans and whites, for example, illustrates a link between a demographic characteristic and mortality levels. Today, African American infant mortality rates remain over twice as high as those for white Americans. The racial difference in infant mortality can be attributed to dissimilarities in the cause of death. African American infants experience a higher rate of infant mortality as a result of their greater likelihood of dying from complications associated with low birth weight and pregnancy, pneumonia, and respiratory distress syndrome.

Even more troubling than the resistance to improvement exhibited by the infant mortality rate is the skyrocketing maternal mortality rate (MMR). As seen in Box 8.9 the US maternal mortality rate was reduced to its lowest level historically around 1990. Since that date, the MMR has steadily risen and in years subsequent to 2020 it has continued to rise (even discounting the impact of the COVID-19 pandemic on maternal health). This increase in maternal mortality is even more striking in that the MMR for comparable countries has continued to fall. This phenomenon represents an example of the influence of various demographic attributes on vital rates, especially since the rise of MMR cannot be attributed to the "usual suspects" (e.g., detrimental lifestyles, poor diet, lack of prenatal care).

**Box 8.9: Infant and Maternal Mortality Rates United States 1920–2020**

| *Year* | *Infant mortality*[a] | *Maternal mortality*[b] |
|---|---|---|
| 2020 | 5.4 | 2.4 |
| 2015 | 5.9 | 1.8 |
| 2010 | 6.1 | 1.3 |
| 2005 | 6.9 | 1.0 |
| 2000 | 6.9 | 1.0 |
| 1990 | 9.2 | 0.8 |
| 1980 | 12.6 | 0.9 |
| 1970 | 20.0 | 2.2 |
| 1960 | 26.0 | 3.7 |
| 1950 | 29.2 | 8.3 |
| 1940 | 47.0 | 37.3 |
| 1930 | 64.6 | 67.3 |
| 1920 | 85.8 | 79.9 |

[a] Deaths per 1000 live births

[b] Deaths per 10,000 live births

*Source* National Center for Health Statistics

## Causes of Death

Significant changes have occurred with regard to the major causes of death affecting US society. Current death patterns will drive present and future medical research in terms of treatment modalities and prevention programs. As long as heart disease and cancer dominate as causes of death in the US and the world, billions of dollars will be invested in reducing them. In turn, medical developments and the advancement of new models that can lead to the reduction of illness will alter the structure of sickness and death. The demand for medical care, drugs and supplies, physical facilities, and health manpower will be shaped by these efforts and the resulting change in sickness and death patterns. To the extent that certain diseases are eliminated, the demand for care related to those diseases (with the possible exception of immunizations) will be eliminated. In regard to the introduction of "new" forms of sickness and death, they too will shape the demand for health services. The rise in obesity, in particular the prevalence of morbid obesity, is driving up the rates of diabetes and heart disease.

***In 2021 gun-related deaths surpassed motor vehicle accidents as the leading cause of death for those under 19 years old for the first time.***

At the macro level, efforts to link environmental factors to death have heightened the population's awareness regarding the presence of carcinogens and other environmental health threats. The market here is both directly related to healthcare (e.g., ongoing research uncovering the connection between pollution and death) and indirectly related (e.g., the development of a medical waste industry).

Box 8.8 compares death rates from various causes over time. Heart related causes and cancer—"modern" "maladies" now account for about 48% of all mortality. The dominant causes of death early in the twentieth century have waned in importance due to improvements in disease prevention and treatment.

**Box 8.10: Proportion of Deaths Attributed to Leading Causes 1900 and 2020**

| *Cause of death* | *1900* | *2020* |
|---|---|---|
| Heart disease | 8.0 | 20.0 |
| Cancer | 3.7 | 17.4 |
| Tuberculosis | 11.3 | * |
| Diarrhea | 8.1 | * |
| Chronic lower respiratory diseases | * | 4.1 |
| Stroke | 4.5 | 4.7 |

| | | |
|---|---|---|
| Liver disease | 5.2 | 1.6 |
| Accidents | 5.1 | 4.9 |
| Alzheimer's disease | * | 3.4 |
| Diabetes | * | 3.0 |
| Pneumonia/influenza | 10.2 | 1.2 |
| Nephritis and related | * | 1.6 |
| Suicide | * | 1.3 |

*Negligible proportion
*Source* National Center for Health Statistics

Although epidemiologists generally attribute deaths to common acute and chronic conditions, there is an argument to be made for the *real* causes of death in contemporary US society. Various researchers have identified the underlying causes for a large proportion of deaths (or the "causes of causes"). As indicated in Box 8.11, diet (coupled with limited exercise) was the underlying cause for the greatest number (20%) of all reported deaths in 2020. Tobacco use accounted for another 14% despite the overall decline in tobacco use.

**Box 8.11: Actual Causes of Death United States 2020**

| *Cause* | *Number* | *Percent (%)* |
|---|---|---|
| Tobacco | 480,000 | 14.2 |
| Diet | 680,000 | 20.1 |
| Alcohol | 140,000 | 4.1 |
| Microbial agents | 35,000 | 1.0 |
| Toxic agents | 100,000 | 2.9 |
| Firearms | 45,222 | 1.3 |
| Sexual behavior | 63,500 | 1.9 |
| Motor vehicles | 46,270 | 1.4 |
| Illicit drug use | 100,000 | 2.9 |
| Total | 1,689,992 | 49.8 |

*Source* Calculated by author

## Sources of Mortality Data

The primary source of mortality data in the United States is the government death registry maintained by the National Center for Health Statistics. This registry is compiled from death certificates filed at the local level (i.e., county health department) which are batched for each state and forwarded to the NCHS for processing and analysis. Information on cause of death in the US is compiled from certificates filed with health authorities on the occasion of any death. Since virtually every death is accompanied by a death certificate, the information on cause of death is fairly complete.

The data collected on the standard death certificate include primary cause of death, contributing causes, and individual demographic and socioeconomic characteristics such as sex, race, ethnicity, last occupation, place of residence, and place of death. Using these data, demographers can begin to study the relationship between the cause of death and a variety of demographic variables.

In addition to the general repository of death data, some cause-specific registry are maintained by NCHS and other departments of the federal government. There are registries maintained on infant deaths and HIV/AIDS deaths for example.

There are several potential problems related to data compiled from death certificates. While identifying the cause of death may seem relatively easy to a layperson, in practice it is often difficult to determine the precise cause of death, potentially resulting in the incorrect assignment of cause. Some deaths are complicated in that more than one condition is present (e.g., cancer and pneumonia) with several bodily systems affected (e.g., heart and lungs). Further, it is often difficult to distinguish between and among the primary and contributing causes, especially since the proximate cause of death may not represent the ultimate explanation.

## Key Points

- As one of the three primary demographic processes an analysis of mortality is considered critical to an understanding of population growth and change
- The most basic way to measure mortality is simply to count the number of deaths.
- Demographers have developed a variety of metrics for use in analyzing mortality patterns
- The crude death rate expresses mortality as the number of deaths per 1000 population
- Age-adjusted death rates provide a more accurate picture of mortality than the crude death rate.
- Infant mortality refers to the number and rate of deaths recorded for infants under one year of age.

- Maternal mortality refers to the number and rate of deaths to mothers as a result of pregnancy or childbirth
- The US infant mortality rate is stubbornly resistant to improvement, while the maternal mortality rate has skyrocketed in recent years
- The leading causes of death provide insights into the nature of mortality, and these have changed significantly over the years
- Calculating potential years of life lost is a useful method for determining the impact of mortality
- Life tables represent a useful device for conceptualizing mortality patterns and are used in a variety of settings for determining future population levels

## Discussion Questions

8.1 How have the primary causes of death changed over the past 100 years? What factors have contributed to these changes? What are the implications for the life expectancy of the US population as a result of these changes?

8.2 What explains the fact that the US maternal mortality rate is increasing, particularly compared to other similar countries? Why is the high observed rate more pronounced among certain minority groups?

## Exercises

*Exercise 8.1: Calculating Mortality Rates*

Using the data provided below, students should calculate the crude death rate, the infant mortality rate, and the maternal mortality rate:

| | | |
|---|---|---|
| Crude death rate: | Population: 100,000 | Deaths 864 |
| Infant mortality rate: | Live births: 5000 | Infant deaths: 50 |
| Maternal mortality rate: | Live births: 5000 | Maternal deaths: 3 |

*Exercise 8.2: Developing a Mortality Profile*

To develop a better understanding of mortality patterns, students should develop a mortality profile for the county of their choice. For this project students should select a county and obtain the following information for that county:

- Identify current death rates for the populations of the respective counties
- Identify historical trends related to mortality for the county

- Identify the leading causes of death for the county's population
- Draw conclusions concerning mortality patterns/trends for the county under study.

## Reference

National Center for Health Statistics. (2006). United States life tables, 2003. *National Vital Statistics Reports, 54*(14).

## Additional Resources

Ezzati, M., Friedman, A. B., Kulkarni, S. C., & Murray, C. J. L. (2008). The reversal of fortunes: Trends in county mortality and cross-county mortality disparities in the United States. *PloS Medicine*. https://doi.org/10.1371/journal.pmed.0050066

Hyder, A. A., Rotllant, G., & R. H. Morrow (1998). Measuring the Burden of Disease: Healthy Life-Years. *Am J Public Health* 88(2): 196–202.

Yusuf, F., Martins, J., & Swanson, D. (2014). *Methods of demographic analysis*. Springer.

CHAPTER 9

# Demographic Processes: Migration

**Abstract** This chapter introduces the process of migration, one of the three basic demographic processes. Over time, migration has become more of a contributor to population change than either fertility or mortality, reflecting the fact that migration has become the most dynamic of the three processes within the US and internationally. The factors that contribute to migration and the implications of migration for both the sending and receiving areas are noted. The characteristics of migrants versus non-migrants are considered and trends in immigration patterns reviewed.

## Introduction

Migration, or geographic mobility, is the third component of population change (along with fertility and mortality). Migration is the most dynamic and complex of the three population processes, as well as the most difficult to measure. While death occurs once for each individual and the average number of births per woman in the United States is less than two, migration is a much more frequent event for most Americans. Recent estimates indicate that the typical American moves 12 times between birth and death. About 10% of the population changes residence each year.[1]

[1] The COVID-19 pandemic had an as-yet-undetermined impact on internal migration patterns. On the one hand, movement within the US was severely restricted during the pandemic. On the other, many households were forced to combine resulting in significant reshuffling of the population. Figures on internal migration during the pandemic years need to be interpreted with caution.





R. K. Thomas, *Demography: An Introduction to Population Studies*, Springer Texts in Social Sciences, https://doi.org/10.1007/978-3-031-56623-3_9

At the state and county levels, the impact of migration is felt more immediately than the effect of fertility or mortality. As birth rates and death rates have fallen, migration has come to play an even more important role in population change. The effects of migration can be significant in the short run for population size and composition at both the point of origin and the point of destination. Persistent long-term migration flows affect subsequent population change in areas receiving migrants, as well as in areas losing residents through out-migration are similarly affected by the process.

## Migration Concepts and Measurement

### *Definitions*

*Migration* refers to a physical move involving an intended permanent change in residence. A residence is defined as the place where a person usually sleeps and eats. Permanent change in residence implies that the person or household in question intends to stay in the new residence for some period of time. Having any residence at all implies some level of permanency in what is recognized as appropriate housing (e.g., apartments and duplexes), although certain categories of individuals do not have recognized residences (e.g., the homeless). Daily or seasonal movement to and from jobs or for climatic reasons does not qualify as migration. Communities such as Daytona Beach, Florida, and Sturgis, South Dakota, for example, experience short-term population increases due to tourism that are not considered to be migration.

***Demographers classify migration into two major categories–international and internal.***

International Migration. Demographers classify migration into two major categories–international and internal. Persons involved in migration either move between countries or they move within the boundaries of a single country. *International migration* refers to the intended permanent movement of people between one country and another. Persons migrating to a country are referred to as *immigrants*, while individuals moving out of a country are labeled *emigrants*.

Every country has laws and policies that govern international migration, especially immigration. In the United States, international migration is regulated by immigration laws that establish the conditions for immigration into the country. These laws establish country-specific limits on the number of persons who may legally move to the United States in any given year as well as the circumstances that justify such a move. Immigration law in the US has limited effect on emigration from this country; most US citizens are free to leave as long as some country will allow entry. Box 9.1 describes the process for immigrating to the US.

**Box 9.1: Categories of Immigrants to the US**
There is a common misperception that the United States has an "open door" policy with regard to immigration. It is believed by many that any foreign national who wants can move into the country. This is clearly not the case, and there are strict rules that govern the admission of legal immigrants into the US. The various pathways to immigration through permanent residency are listed below, along with the most recent figures for the volume generated by each category:

| *Category* | *Number* | *Percent* |
|---|---|---|
| Family-based visas | 709,904 | 69 |
| Immediate relatives (reunification) | 505,765 | 49 |
| Family-sponsored | 204,139 | 20 |
| Employment-based visas | 139,458 | 14 |
| Diversity-based visas | 43,463 | 4 |
| Refugees/asylum seekers | 106,911 | 10 |
| All other categories | 32,039 | 3 |

*2019 data were utilized although later statistics are available. Figures generated during the COVID-19 pandemic (i.e., 2020–2022) are not reflective of the figures for migration in typical years
*Source* US Department of Homeland Security (2022)

Internal Migration. Internal migration refers to a change of residence within a particular country. Internal migration is generally less regulated (and measured) than international migration. Demographers refer to internal migrants coming into an area as *in-migrants*, while those leaving an area are termed *out-migrants*. Some countries do maintain registries of their citizens that allow for the tracking of internal migration. Within the United States, internal migration is basically unimpeded.

***Internal migration refers to a change of residence within a particular country.***

Internal migration can be categorized as either short-distance or long-distance, and a hierarchy of definitions has been created to reflect the distance of the move. Anyone who permanently changes residence (regardless of distance) is classified as a *mover*. In order for a mover to be designated as a *migrant* in the US, the mover has to change his or her county of residence. The county was chosen for the mover/migrant distinction because it was felt that movement across such a boundary involves a substantial change in social and economic circumstances. From this perspective, all migrants are movers, but a mover is not necessarily a migrant. Other useful distinctions also reflect the distance and nature of the movement. *Intrastate* migration refers to movement within a state, for example, while *interstate* migration refers to movement between two states.

*Migration streams*, or the flow of relatively large numbers of persons from one area to another are a common phenomenon. These streams involve large numbers of migrants moving from one location to another for the same reason. For example, the flow of African Americans from the South to the Northeast and Midwest during the 1930s and 1940s constituted a migration stream. More recently, African American migration streams from the Midwest and Northeast to the South have been identified. Thus, some African Americans were born in the South, resided in the North for some length of time, and then they or their descendants returned to the South at a later point in time.

***Migration streams, or the flow of relatively large numbers of persons from one geographic area to another, are a common phenomenon.***

Similarly, international migration patterns often exhibit migration streams. While we often think of a move of such significance as an individual decision, most international migration is part of mass movements that include large numbers of migrants moving from one country to another.

### *Migration Attributes*

Migration can be classified as voluntary or involuntary. Voluntary migration occurs at the migrant's volition and usually involves moves for economic needs, retirement, family reasons or simply for a change of scene. Involuntary migration is typically a result of political or religious persecution, wars, civil unrest, or famines and other natural disasters. The discussion in this chapter will focus primarily on voluntary migration.

A distinction should also be made between legal and illegal migration. Legal immigration refers to those entering a country with the formal permission of that country. In the US, this typically involves the acquisition of a "green card" that allows for the permanent or fixed-period residence of an immigrant to this country. Immigrants to any county must possess a current, legitimate passport before entry.[2] Other countries have similar legal requirements.

Illegal immigrants are those who enter a country without proper legal authorization. Many of these are temporary movers seeking short-term employment or visits with family members. Others enter the country illegally with the intent of staying permanently. While accurate records are maintained by federal immigration authorities (Immigration and Customs Enforcement) on legal immigration, limited data are available on illegal immigrants.

[2] Asylum seekers and refugees seeking entrance to the US may not have the documentation needed by other categories of immigrants.

## Migration Measurement

The measurement of internal migration in the US is problematic because there is no migration registry or other mechanism to track the movement of the population. Migration data are most often derived by comparing addresses at two points in time based on a survey or some type of systematic record keeping. This approach, however, does not consider the number or nature of the moves that may have occurred between the two dates specified.

The migration concept is difficult to apply to certain categories of people in transit (e.g., migrant workers and "snowbirds") for whom the move is not expected to be permanent. Moreover, the growing complexity of living arrangements reflected in non-traditional households and blended families makes the measurement of migration even more challenging.

Demographers have developed a number of measures for quantifying migration, but all face difficulties related to the concepts employed and the data sources available. Accurate records are maintained in the US for international migration (i.e., for immigrants and emigrants). However, these records pertain primarily to legal immigrants; statistics on the growing number of illegal immigrants are based on estimates and are understandably less accurate. Since no formal records are maintained on internal migration in the US, we are left with a significant gap in our understanding of the year-to-year mobility of our population.

***The most common measure of migration is simply a count of the number of individuals moving from one geographic area to another.***

The most common measure of migration is simply a count of the number of individuals moving from one geographic area to another. The number of people moving into a county during a given year and the number of people moving out of that county, for example, constitute crude indicators of migration. Since they represent raw numbers unrelated to any population characteristics, a variety of rates have been developed for measuring the flow of migrants.

While the comparison of the absolute numbers of movers, non-movers, and migrants is important, rates need to be calculated when there are size differences between the areas being compared. Rates for in-migration, out-migration, net migration and gross migration are typically generated by demographers. The numerator for these rates is the total for each migration category (e.g., in-migrants, out-migrants); the denominator depends on what is considered as the population at risk for migration. The identification of the population at risk, or persons with at least some probability of moving over a given timeframe, is complicated because each rate is based on a different risk group.

*Net migration* is a measure of the absolute difference between in-migration and out-migration for an area over a given time period. For example, if county X in state Y gained 10,000 in-migrants and lost 5000 out-migrants between 2005 and 2010, the figure for net migration would be 5000. (Had the figures been reversed, net migration would have equaled – 5000.) *Gross migration* is used to measure the total amount of migration by adding in-migrants to out-migrants. Thus, gross migration for county X for this time period would be 15,000. Gross migration is used to assess the total amount of population turnover, while net migration is a better indicator of the outcome of the migration process. Box 9.2 provides a depiction of the various categories of migrants.

**Box 9.2: Categories of Migrants**

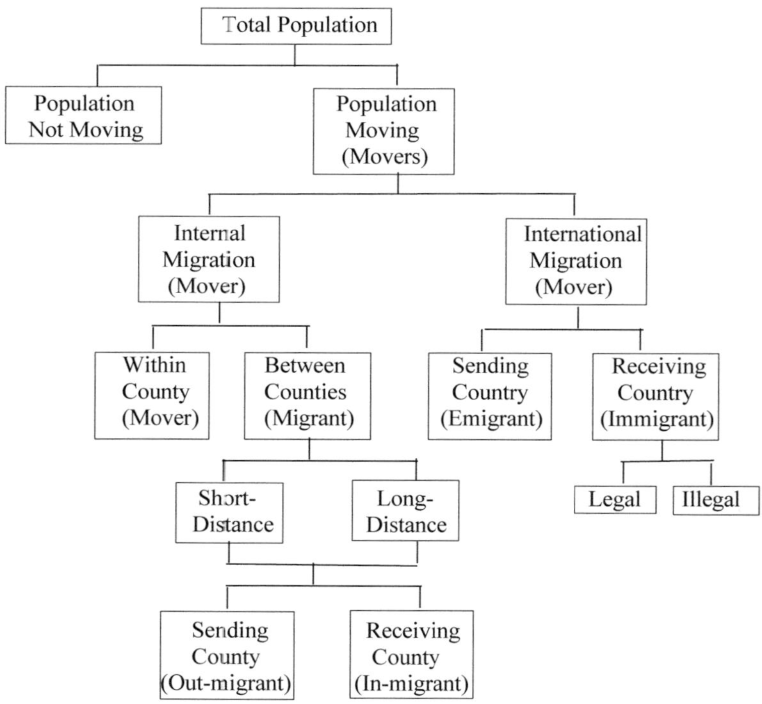


To illustrate the issues with calculating in- and out-migration rates, consider the out-migration rate for a specific city in a one-year period. The numerator for the out-migration rate is the number of out-migrants, while the denominator is the population at the beginning or in the middle of the one-year period. This calculation is straightforward, but identifying the population at risk for the in-migration rate is problematic, since virtually the entire population of the United States could be considered at some risk of moving into the city. For this reason, the denominator for the in-migration rate is typically the same one used for calculating the out-migration rate—that is, the total population of the city in question. This means that the rate generated is technically the percentage of population increase due to in-migration. (see Box 9.3 for migration rate calculations.)

The rate of migration has a substantial impact on the population size and characteristics of the affected communities. An examination of migration trends for Florida illustrates this point. The population of Florida grew substantially (17.6%) between 2000 and 2010, although that growth rate was considerably below those seen in previous decades. Net migration—not natural increase—has historically accounted for over 85% of the state's growth.

**Box 9.3: The Calculation of Migration Rates**

Migration rates are relatively easy to calculate if the required data are available. Data (numerators) on international migration are available from Immigration and Customs Enforcement and internal data are available from IRS records and sample surveys. Population figures (denominators) can be drawn from Census Bureau counts or from estimates generated by other sources. These basic rates can be adjusted to reflect other factors such as age and marital status as desired.

$$\text{In-migration rate} = \frac{\text{Persons moving into area in } Y_1}{\text{Midpoint population estimate in area for } Y_1} \times 1000$$

$$\text{Out-migration rate} = \frac{\text{Persons moving out of area in } Y_1}{\text{Midpoint population estimate in area for } Y_1} \times 1000$$

$$\text{Net-migration rate} = \frac{\text{Persons moving into area in } Y_1 - \text{Persons moving out of the area in } Y_1}{\text{Midpoint population estimate in area for } Y_1} \times 1000$$

$$\text{Gross migration rate} = \frac{\text{Persons moving into area in } Y_1 + \text{Persons moving out of the area in } Y_1}{\text{Midpoint population estimate in area for } Y_1} \times 1000$$

$$\text{Migration efficiency} = \frac{\text{Net migrants (in-migrants} - \text{out-migrants)}}{\text{Gross migrants (in-migrants} + \text{out-migrants)}}$$

For a hypothetical population of 10,000 that experienced during $Y_1$ 2000 in-migrants and 1000 out-migrants the following rates would be generated:

In-migration rate: 200 per 1000 population
Out-migration rate: 100 per 1000 population
Net migration rate: 100 (a net gain) per 1000 population
Gross migration rate: 300 (moves in or out) per 1000 population
Migration efficiency: 33%.

## Characteristics of Migrants

### *International Migrants*

The volume of immigrants entering the US at any point reflects a number of factors. Chief among these are the immigration laws that control the flow of immigrants into the country. Contrary to popular belief, someone in another country cannot simply decide they want to move to America. There are a limited number of pathways for accomplishing this. Some immigrants, of course, find various ways to skirt these laws, and this has resulted in a significant number of "undocumented" immigrants residing in the US.

***The two factors affecting the level of immigration are conditions "on the ground" in the sending country and the immigration policies of the receiving country.***

The other factor affecting immigration to the US (and any other country for that matter) is the conditions affecting populations in other parts of the world. Many immigrants may be voluntarily seeking a better life or safer environment. Others may be forced to immigrate due to natural disasters, war, or persecution. The tens of millions of displaced persons in displaced persons "camps" around the world attests to the level of involuntary migration that currently exists.

In order to understand the impact of immigration on American society, one must understand the nature of the immigrants themselves. One million or more immigrants entering the US every year has important implications for the nation's social, economic and political characteristics. Further, the 14% of US residents who are foreign-born cannot help but influence the nature of our society.

The region of origin is an important consideration in the analysis of immigration trends. In 2021 of the estimated 45.5 immigrants in the US the majority (49.7%) were from Latin America. This was followed by 30.7% from Asia and 11% from Europe. Africa accounted for 5.7% and Oceania for less than one percent.

Immigrants to the US today originate in two primary world regions. Latin America accounted for 42% of immigrants between 2011 and 2020, while Asia recorded 39%. The other one-fifth of immigrants originated in Africa (10%), Europe (8%) and all other regions (1%).

When these figures are broken down by country of origin, five countries accounted for nearly half of the immigrants in 2020. These were Mexico, India, China, the Philippines and El Salvador. Box 9.4 depicts the distribution of legal immigrants by country of origin in 2020.

**Box 9.4: Country of Origin for Immigrants United States 2020**

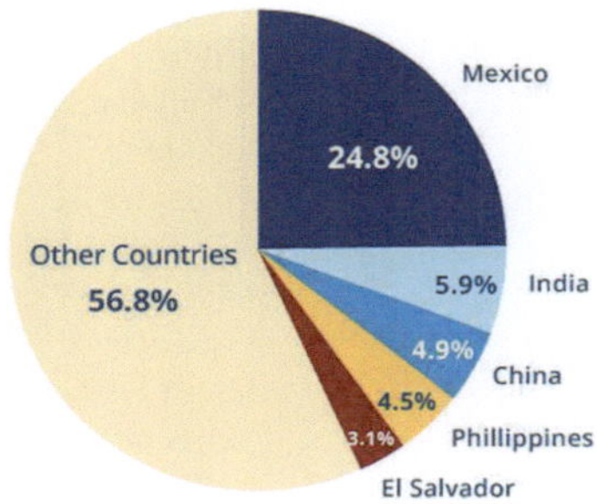


*Source* US Census Bureau

As is the case with most demographic variables, immigrants are not randomly distributed across the US. When immigration trends are analyzed at the subnational level it is found that immigrant destination is skewed toward relatively few states. Immigrants are predominantly found in nine states (listed in order of number): California, Texas, New York, Florida, New Jersey, Illinois, Massachusetts, Washington, and Virginia. The combined immigrant population of these seven states accounted for 60% of the total foreign-born population as of 2021. Many patterns of migrant distribution have remained fairly constant over time. However, there has been a steady dispersion of immigrants across the US—to the point that by 2010 every single US county had some immigrants in residence. Box 9.5 depicts the distribution of immigrants among the various states in 2021 and Box 9.6 depicts the distribution of foreign-born residents across the country.

**Box 9.5: Distribution of Immigrants and Share of Population United States 2021**

| | *Number* | *Percent (%)* | | *Number* | *Percent (%)* |
|---|---|---|---|---|---|
| Alabama | 177,000 | 3.5 | Montana | 24,000 | 2.2 |
| Alaska | 59,000 | 8.1 | Nebraska | 144,000 | 7.4 |
| Arizona | 916,000 | 12.6 | Nevada | 580,000 | 18.4 |
| Arkansas | 143,000 | 4.7 | New Hampshire | 82,000 | 5.9 |
| California | 10,452,000 | 26.6 | New Jersey | 2,135,000 | 23.0 |
| Colorado | 570,000 | 9.8 | New Mexico | 193,000 | 9.1 |
| Connecticut | 549,000 | 15.2 | New York | 4,427,000 | 22.3 |
| Delaware | 101,000 | 10.1 | North Carolina | 870,000 | 8.2 |

| | | | | | |
|---|---|---|---|---|---|
| District of Columbia | 89,000 | 13.3 | North Dakota | 34,000 | 4.4 |
| Florida | 4609 | 21.2 | Ohio | 585,000 | 5.0 |
| Georgia | 1083 | 10.0 | Oklahoma | 231,000 | 5.8 |
| Hawaii | 27,000 | 18.8 | Oregon | 412,000 | 9.7 |
| Idaho | 115,000 | 6.2 | Pennsylvania | 939,000 | 7.2 |
| Illinois | 1,805,000 | 14.2 | Rhode Island | 159,000 | 14.5 |
| Indiana | 379,000 | 5.6 | South Carolina | 271,000 | 5.2 |
| Iowa | 173,000 | 5.4 | South Dakota | 31,000 | 3.5 |
| Kansas | 202,000 | 6.9 | Tennessee | 371,000 | 5.3 |
| Kentucky | 182,000 | 4.0 | Texas | 5,092,000 | 17.2 |
| Louisiana | 200,000 | 4.3 | Utah | 278,000 | 8.3 |
| Maine | 56,000 | 4.1 | Vermont | 27,000 | 4.2 |
| Maryland | 979,000 | 15.9 | Virginia | 1,070,000 | 12.4 |
| Massachusetts | 1,227,000 | 17.6 | Washington | 1,143,000 | 14.8 |
| Michigan | 683,000 | 6.8 | West Virginia | 28,000 | 1.6 |
| Minnesota | 487,000 | 8.5 | Wisconsin | 301,000 | 5.1 |
| Mississippi | 63,000 | 2.1 | Wyoming | 20,000 | 3.4 |
| Missouri | 251,000 | 4.1 | | | |

*Source* US Census Bureau

**Box 9.6: Distribution of Recent Immigrants United States 2011–2015**

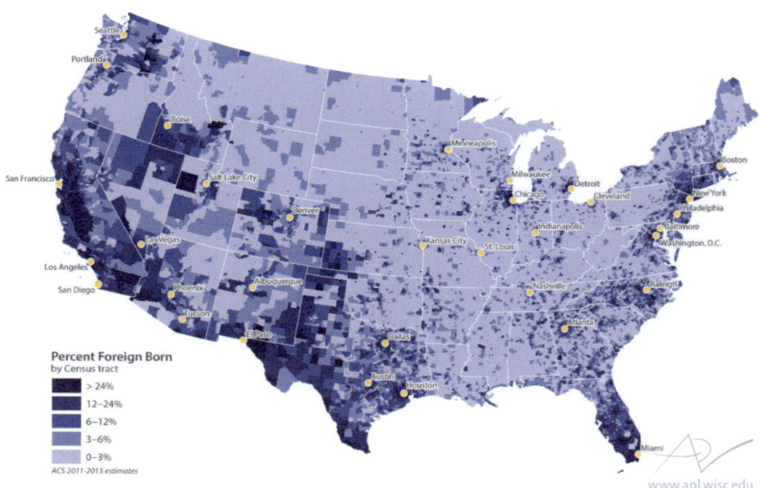


*Source* US Census Bureau

The top 10 metropolitan areas (MSAs) in terms of immigrant settlement are presented in Box 9.7. The New York metropolitan area was by far the greatest recipient of newly arrived immigrants. Los Angeles and Miami were the second most popular areas of settlement for immigrants (with Miami accounting for 29% of the most recent immigrants). These 10 urban areas combined accounted for 48% of immigrants obtaining permanent resident status.

The geographical regions where immigrants settle have also changed. According to census data, Boston Seattle and Dallas moved into the top ten locations for new immigrants when they arrive in the US. San Jose—part of Silicon Valley—recently moved ahead of Miami into the number one destination for recent immigrants.

**Box 9.7: Distribution of New Immigrants Major Metropolitan Areas 2019[a]**

| *Metropolitan area* | *Number* | *Percent* |
|---|---|---|
| New York | 155,187 | 15 |
| Los Angeles | 75,859 | 7 |
| Miami | 73,040 | 7 |
| Houston | 40,137 | 4 |
| Washington, DC | 33,964 | 3 |
| Dallas | 32,984 | 3 |
| Chicago | 34,881 | 3 |
| San Francisco | 32,787 | 3 |
| Boston | 24,717 | 2 |
| Seattle | 19,850 | 1 |

[a] 2019 data were utilized although later statistics are available. Figures generated during the COVID-19 pandemic (i.e., 2020 and 2021) are not reflective of typical years of immigration data

*Source* US Department of Homeland Security (2022)

Much has been made of the differences between immigrants and native-born Americans with regard to their demographic and socioeconomic characteristics. International migrants tend to be younger that the native born and, in the US today, 55% of immigrants are working age (25–54) compared to 33% of the native-born. While the proportion of native-born Americans who are "old" (65 years and older) is around 16% and continuing to increase, there is a negligible proportion of immigrants who are seniors. Although in past years the majority of immigrants were single males, this situation has changed significantly over time with female immigrants and/or family-unit immigrants becoming increasingly common. In fact, in 2020 64% of immigrants were in a married-couple family, compared to 58% of native-born Americans. Despite the perception of the uneducated immigrant, it is found that immigrants overall are only slightly less educated than native-born citizens. At the same time, a higher proportion of immigrants are in the workforce and employed than non-immigrants. Box 9.8 compares the occupational categories of immigrant and non-immigrant workers in the US labor force.

**Box 9.8: Occupational Distribution of Immigrant and US-born Workers United States 2020**

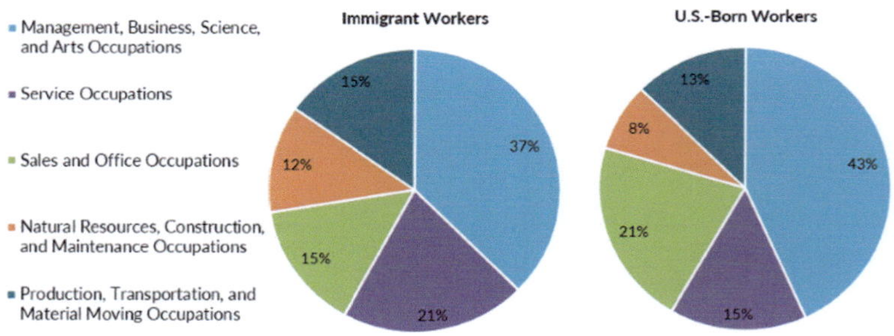


*Source* US Census Bureau

### *Internal Migrants*

Given the fact that approximately 33 million Americans move every year (and sometimes more than once), it is important to understand the characteristics of this significant segment of the population. About 10% of the population changes residence each year. The typical American today moves 12 times between birth and death.

Based on data from 2018–2019 it is found that, of the 10% of Americans that moved, 65% were short-distance movers, remaining in the same county and not qualifying as migrants. Some 17% moved to a different county in the same state, while only 14% moved to a different county in a different state. Many, if not most, of those moving to a different county in the same state or moving to another state would be considered long-distance movers as well as migrants.

Internal migrants tend to be younger, with 25% of those 18–24 moving between 2017 and 2018. This compares to only 6% of senior citizens. Not surprisingly, renters are more likely to move than homeowners (24% vs. 6%). Low-income residents are also more likely to move, with 14% of those in the bottom income quartile and 11% in the top income quartile moving during this period.

***Given the fact that approximately 33 million Americans move every year, it is important to understand their characteristics.***

## Migration Implications for US Society

### *International Migration*

Immigrants to the US have had an indelible impact on American society, with implications for its economy, its social structure and its culture. Without post-World War II immigration, the population of the United States would be declining in number and much older in age structure. We are once again a nation of immigrants with one in eight Americans being foreign born and more than half of America's school children being part of an immigrant minority group.

Immigrants have dominated certain occupations with the areas of employment varying over time and by country of origin. The dominant immigrant group, Hispanics, has long been prominent in the construction, building maintenance and landscaping fields. Hispanics account for 78% of the agricultural labor force. Young immigrants have historically filled many entry-level jobs, and the impact of the decrease in this labor force was noteworthy during the Trump years and the period following the pandemic. Employers found it difficult to fill entry-level jobs and positions in many occupations where immigrants were previously dominant. It is argued that immigrants have taken the jobs that Americans would not take, and their absence has been noteworthy during the period of "full employment" following the COVID-19 pandemic.

Immigrants have long contributed to advances in American culture, and the contributions that have been made by immigrants are almost possible to recount. The cultural mosaic that is America is a result of the "melting pot" effect of generations of immigrants. Their contributions to pop culture are extensive, as evidenced by the popularity of Latin music and the success of the Latin Music Awards. Perhaps the true measure of immigrant influence is the fact that for the past 20 years, salsa has outsold ketchup as the most popular condiment.

***The cultural mosaic that is America is a result of the "melting pot" effect over generations of immigrants.***

There are those who contend that immigration is a negative phenomenon with numerous deleterious effects on American society. It is argued that immigrants take American jobs, do not pay taxes, use an inordinate share of public services, abuse the healthcare system and exhibit a high crime rate. None of the assertions have any merit and in virtually every case it can be demonstrated that the record for native-born Americans is worse. Study after study has demonstrated the positive benefits of high rates of immigration with economists being in the forefront of researchers who point to the positive contributions of immigrants to the economy.

From a demographic perspective, it is important to note that, without the level of immigration of recent years, the US population would not have experienced any growth. Without immigrants and their children, the US population would have actually declined. This is significant not only from a population growth perspective but from an economic perspective. Most of the US economy is built on consumer spending; without the contributions of immigrants to the economy, there would be little of the expansion necessary for economic growth.

### *Internal Migration*

The history of the United States is, to a great extent, a chronicle of migration. While space does not allow all of the details of historical internal migration to be presented, some key trends should be noted. The general flow of population in the US since the colonization of the New World by Europeans has been from east to west. The county's population center at the time of the first census in 1790 was on the east coast; today, the population center is in Missouri. As the country became fully settled, the general trend toward east-to-west movement continued but a more complicated picture emerged. (See the map depicting changes in the US center of population in Box 3.12.)

The level of internal migration that has historically characterized the US population has generally represented a positive force. The high level of geographic mobility has been associated with a high level of economic mobility as change in residence often meant an increase in socioeconomic status. In fact, during the surging economy of the second half of the twentieth century, the maxim was: Every [job] transfer means a promotion. Indeed, recent declines in economic expansion are blamed as least partly on the fact that Americans have become less mobile.

In the past it was argued that the mobility characterizing the American population was the "great leveler" on the assumption that as people migrated to other parts of the country, the US population would become more homogenous. Although there is some validity to this argument, the surprise for prognosticators has been the degree to which the population has not become homogenous. In fact, there is the current argument that regional differences have actually been augmented as an unintended consequence of the high level of geographic mobility.

***The high level of geographic mobility has been associated with a high level of economic mobility.***

## Trends in Migration

As previously noted, migration has become a major determinant of population change in the US and around the world. Changes in the number, characteristics and location of immigrants (both internal and international) represent the most dynamic aspects of population change. The US has always been a nation of movers, but this movement has taken on new meaning with the massive influx of foreign nationals in recent years.

### *International Migration*

The volume and nature of immigration to the United States have varied greatly over the history of the country. Box 9.9 provides the decade-by-decade record of immigrant flows, along with the percentage of total decade population growth accounted for by immigrants. A comparison of the data by decade shows a low of 528,000 immigrants for the period 1931–1940 and a high of 10 million immigrants during each of the first two decades of the 2000s.

Until recently, the contribution of immigration to population growth was greatest during the first two decades of the twentieth century. This contribution declined markedly in the 1930s, and it was only in the 1980s that immigration's contribution began to increase. The recent trend for international migration has been upward, with immigrants accounting for over 40% of population growth in the first two decades of the twenty-first century. This trend has led to an increase in the number of legal immigrants living in the US, from 24 million in 1995 to 45.5 million in 2020, more than tripling the figure for 1970. This figure does not take into consideration illegal immigrants which are estimated at 14.5 million in 2020. Immigrants and their children account for 26% of the 2020 US population.

Up until 1920, the vast majority of immigrants were from Europe. Since 1920 most of the share lost by Europe has been gained by Asia, and in the decade of the 1980s nearly half of all immigrants originated in Asia. Altogether, 85% of all immigrants came from Asia or other countries in North and South America in the 1990s. In 2010, the major sources of legal immigrants by country were Mexico (139,120), China (70,863), India (69,162), the Philippines (58,173), and the Dominican Republic (53,870). These same countries (except for the Dominican Republic) account for the largest numbers of foreign-born currently living in the US. As a result of this shift in country-of-origin, the immigrant population is quite different in culture and language from the majority European-origin population associated with historical immigration patterns. Box 9.9 depicts the decade-to-decade impact of immigration on the US population while Box 9.10 provides the same information broken down by continent of origin.

The large numbers of immigrants from Latin America and Asia in recent decades represent a sharp turnaround from the mid-1900s, when immigration largely came from Europe. In the 1960s, no single country accounted for more than 15% of the US immigrant population, but Italians were the top origin group, making up 13% of the foreign born in 1960, followed by Germans and Canadians (about 10% each).

Immigrants from Mexico have been the most numerous since 1980, but the composition of new arrivals has changed since 2010. Now, immigrants are more likely to come from Asia, especially India and China. In fact, these two nations displaced Mexico as the top origin countries for new arrivals from 2013 to 2021, but amid the pandemic and related mobility restrictions Mexico has regained its position as the origin of most new arrivals.

**Box 9.9: Immigrants and the Proportion of Population Growth Due to Immigration by Decade for the United States 1831–2020**

| *Decade* | *Immigrants (in thousands)* | *Population growth for decade (in thousands)* | *Percent of population growth due to immigration* |
|---|---|---|---|
| 2011–2020 | 10,298 | 22,700 | 48.1 |
| 2001–2010 | 10,501 | 27,300 | 38.1 |
| 1991–2000 | 9080 | 32,712 | 27.8 |
| 1981–1990 | 7258 | 22,164 | 32.7 |
| 1971–1980 | 4493 | 23,244 | 19.3 |
| 1961–1970 | 3322 | 23,979 | 13.9 |
| 1951–1960 | 2515 | 27,767 | 9.1 |
| 1941–1950 | 1035 | 19,028 | 5.4 |
| 1931–1940 | 528 | 8894 | 5.9 |
| 1921–1930 | 4107 | 17,064 | 24 |
| 1911–1920 | 5736 | 13,738 | 41.8 |
| 1901–1910 | 8795 | 15,978 | 55.0 |
| 1891–1900 | 3688 | 13,047 | 28.3 |
| 1881–1890 | 5247 | 12,792 | 41.0 |
| 1871–1880 | 2812 | 10,337 | 27.2 |
| 1861–1870 | 2315 | 8375 | 27.6 |
| 1851–1860 | 2598 | 8251 | 31.5 |

*Source* US Census Bureau

**Box 9.10: Immigrants by Continent of Origin United States 1831–2020**

| *Decade* | *Total Immigrants*[a] | *Europe* | *Asia* | *Other America* | *Africa* |
|---|---|---|---|---|---|
| 2011–2020 | 10,298 | 836 (8) | 4053 (39) | 4304 (42) | 1042 (10) |
| 2001–2010 | 10,501 | 1266 (12) | 3785 (36) | 4510 (43) | 860 (8) |
| 1991–2000 | 9080 | 1309 (14) | 2890 (31) | 4449 (49) | 382 (4) |
| 1981–1990 | 7256 | 706 (10) | 2814 (39) | 3581 (49) | 192 (3) |
| 1971–1980 | 4493 | 801 (18) | 1634 (36) | 1929 (43) | 92 (2) |
| 1961–1970 | 3322 | 1239 (37) | 445 (13) | 1579 (48) | 39 (1) |
| 1951–1960 | 2516 | 1492 (53) | 157 (6) | 841 (33) | 17 (1) |
| 1941–1950 | 1035 | 622 (60) | 59 (3) | 355 (34) | 7 (0) |
| 1931–1940 | 52,834 | 8 (66) | 15 (3) | 160 (30) | 2 (0) |
| 1921–1930 | 4107 | 2478 (60) | 97 (2) | 1517 (37) | 6 (0) |
| 1911–1920 | 5736 | 4377 (76) | 193 (3) | 1144 (20) | 8 (0) |
| 1901–1910 | 8795 | 8136 (93) | 244 (3) | 362 (4) | 7 (0) |
| 1891–1900 | 3688 | 3559 (97) | 71 (2) | 39 (1) | 1 (0) |
| 1881–1890 | 5247 | 4722 (90) | 68 (1) | 426 (8) | [b](0) |
| 1871–1880 | 2812 | 2262 (80) | 124 (4) | 404(14) | [b](0) |
| 1861–1870 | 2315 | 2064 (89) | 65 (3) | 167 (7) | [b](0) |

[a] Number in thousands, percentages in parentheses
[b] Less than 1000
*Source* US Census Bureau

In terms of the total size, the immigrant population from India increased by 929,000 and from China/Hong Kong by 572,000 between 2010 and 2021. In contrast, the number of Mexican immigrants in the United States declined by more than 1 million during the same period, representing the largest absolute decline of all immigrant groups. The number of immigrants from the Dominican Republic, Venezuela, Guatemala, Honduras, Brazil, Nigeria, Colombia, the Philippines, and El Salvador has increased by at least 200,000 each since 2010.

The states with the largest *numerical* increases between 2010 and 2020 were Florida (up 721,298); Texas (up 712,109); California (up 502,985); New York (up 242,769); New Jersey (up 210,481); Washington (up 173,891); Massachusetts (up 172,908); Pennsylvania (up 154,701); Virginia (up 151,251); Maryland (up 124,241); Georgia (123,009); Michigan (up 116,059); North Carolina (up 110,279); and Minnesota (up 107,760).

The states with the largest percentage increases since 2010 are North Dakota (up 87%); Delaware (up 37%); West Virginia (up 33%); South Dakota (up 32%); Wyoming (up 30%); Minnesota (up 28%); Nebraska (up 28%); Pennsylvania (up 21%); Utah (up 21%); and Tennessee, Kentucky, Michigan, Florida, Washington, and Iowa (all up 20%).

## *Internal Migration*

The history of the United States is, to a great extent, a chronicle of migration. The general flow of population in the US since the colonization of the New World by Europeans has been from east to west. The county's population center at the time of the first census in 1790 was on the east coast; today, the population center is in Missouri. As the country became fully settled, the general trend toward east-to-west movement continued but a more complicated picture emerged. (See the map depicting changes in the US center of population in Box 3.17.)

Rural to urban migration, which began in some areas of the nation as early as 1850, has altered the course of industrial development. The east-to-west and city-to-suburb movement of the population has forever changed the social, economic and political structure of the United States. During the first half of the twentieth century a large number of people left the South, destined primarily for the Midwest and the Northeast but also contributing to the influx of new residents into the West region. This flow primarily involved African Americans who had been displaced from farm work with the mechanization of agriculture. Since the 1950s the outflow from the South has been reversed, and today the South has more population than any of the other three regions. Box 9.11 presents data on migration trends for the 2010–2011 period. In more recent years the migration of persons from the Snowbelt to the Sunbelt has markedly affected both the place of origin and the place of destination.

**Box 9.11: Regional Migration Flows United States c. 2010**

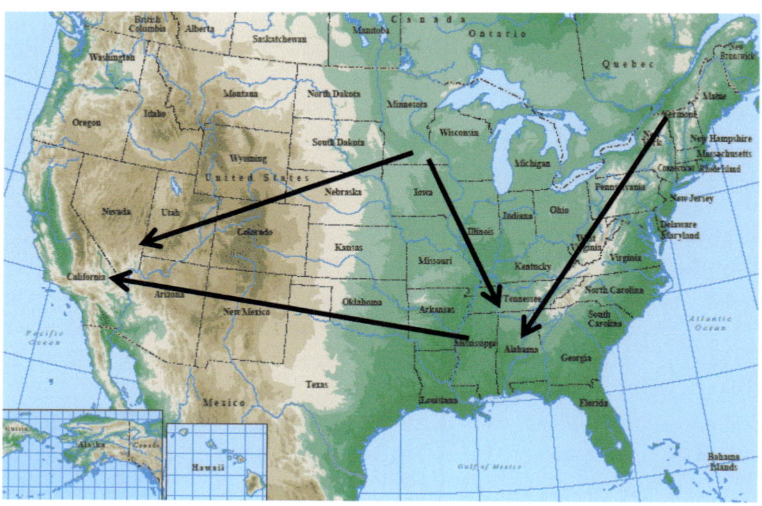

During the last quarter of the twentieth century, new migration patterns were established. From that time on, the South region has been the major beneficiary of internal migration. This is followed by the West region. On the other hand, the Northeast and Midwest regions have suffered from outmigration. In the second decade of the twenty-first century, the West began losing more residents than it gained, leaving the South as the only region with positive net migration. Box 9.12 describes recent patterns of internal migration.

**Box 9.12: Regional Population Change Due to Migration United States 2021[a]**

| | *Northeast* | *Midwest* | *South* | *West* |
|---|---|---|---|---|
| Internal migration | – 227 | – 134 | 153 | 108 |
| Movers from abroad | 145 | 63 | 323 | 144 |
| Net migration | – 146 | – 71 | 576 | 252 |

[a] Figures stated in thousands
*Source* US Census Bureau

There was substantial movement of persons during 2021 (despite the impact of the COVID-19 pandemic), resulting in a net inflow for the South region and a net outflow for the Northeast, Midwest, and West. However, the net flows only tell part of the story. While the West had a net loss of 31,000 persons to the South, 380,000 persons moved from the South to the Northeast. Though the net flows have important implications for demographic change, gross migration is a better measure of population turnover than net migration. The net gain or loss of 25,000 persons for a region over a 5-year period may seem like a small number, but if those who left the area were demographically very different from those who entered it, the impact could be substantial.

The other major trend since World War II has been the shift in residence based on type of community. At the time of the first census in 1790, 95% of the population lived in what today are classified as "rural" areas. With the advent of industrialization in the US, a mass movement from rural areas to urban areas occurred, to the point that today only 16% of US citizens live in rural communities. The rural-to-urban migration trend peaked in the 1970s and a new flow emerged—from urban areas to suburban areas. Since that period the major flow has been out of the nation's central cities and into surrounding suburbs. Today, more Americans live in communities classified as suburbs than in any other type of community.

In recent years two additional migration streams have been identified. One is the movement from the suburbs to the exurbs. Exurbs refer to communities beyond the boundaries of a metropolitan area but in an adjacent county. As suburban areas have become more crowded and their character has begun to change and jobs have become more decentralized, some are opting for residence at ever greater distances from the central city. The second trend involves the "reurbanization" movement. A number of factors have been driving a shift of population back into the inner city in many places. A new generation that is not comfortable with the homogenous nature of the suburbs and rising transportation costs were lured by downtown redevelopment initiatives back into urban areas. Some have referred to this trend as the "gentrification" of the inner city, as affluent individuals and families move into areas in decline, causing the cost of housing to increase and often displacing the original residents. Due to the influence of the COVID-19 pandemic another trend has emerged driven by the increasing number of remote workers, a trend with implications for historical settlement patterns.

***A major trend since World War II has been the shift in residence based on type of community.***

Additional insight into internal migration can be gained by examining the respective characteristics of movers and non-movers. Movers are considerably younger than non-movers, recording a median age nine years less. The youth of movers is reflected in the concentration of persons under 35. Because of the preponderance of youthful movers, areas receiving migrants gain a younger population, in general, while areas losing migrants "age" more rapidly because of the loss of younger persons. Continued gain or loss can have a significant impact on both the size and age structure of populations sending and receiving migrants. One notable exception to the youth selectivity of migration is the migration in the movement of older persons to certain retirement areas of the United States. Even so, the overall proportion of persons above the age of 55 who move is low.

Overall, domestic movers are slightly more likely to be male and considerably more likely to be never married and better educated. The incomes of movers, however, tend to be somewhat lower on average and the poverty rate somewhat higher.

## Sources of Migration Data

Data on migration within the United States are derived from censuses, surveys, and administrative records. On Census Bureau forms respondents are typically asked if they lived in the same house one year ago or five years ago. The American Community Survey is the most useful and current source of this information. On other federal surveys, respondents are asked how long they have lived at their current location and how many times they have moved in the last three years.

The other sources of internal migration data, administrative registries such as Social Security and the Internal Revenue Service, use a two-points-in-time comparison of addresses to generate data on the volume and nature of residential moves. Data available from the IRS provide the number of household members (and their incomes) moving from one county to another from one year to another. This database indicates how many people (along with their associated incomes) moved from County A to County B between, for example, 2010 and 2011. The IRS data come closest to complete coverage of the population but none of these data sources captures the complexity of contemporary migration patterns.

More accurate and detailed data are available on international migration. As with internal migration, the American Community Survey elicits information on international migration (presumably both legal and illegal). However, the best data on legal migration are provided by the agencies charged with monitoring ingress and egress into and out of the country. The Immigration and Customs Enforcement agency (ICE), now under the Department of Homeland Security, maintains records on immigration and emigration and regularly publishes this information in its *Yearbook of Immigration Statistics*. This agency also generates annual estimates of the volume of illegal immigration into the US. The US Department of State is charged with tracking immigrant and non-immigrant visas issued and reports that information annually.

## Key Points

- Demographers classify migration into two major categories—international and internal.
- International migration refers to the intended permanent movement of people between one country and another.
- Internal migration refers to a change of residence within a particular country.
- Migration streams, or the flow of relatively large numbers of persons from one geographic area to another, are a common phenomenon.
- The most common measure of migration is simply a count of the number of individuals moving from one geographic area to another.
- The two primary factors that influence international migration are the circumstances "on the ground" in the sending country and the immigration policies of the receiving country.
- The cultural mosaic that is America is a result of the "melting pot" effect over generations of immigrants.
- Today immigrants from Latin America and Asia account for the lion's share of US immigrants.
- Americans remain a relatively mobile population, although the level of geographic mobility has declined in recent years.
- The high level of geographic mobility has historically been associated with a high level of social mobility.

- The South region has become the primary destination for internal migrants in the US, while the Northeast and Midwest continue to lose population.
- A major trend since WWII has been shifts in residence in the US based on type of community.

## Discussion Questions

9.1 The issue of immigration into the United States is a controversial one. In your opinion is the current level of immigration at good thing or a bad thing? Explain your response.

9.2 Some people argue that the US should close its borders to any more immigrants. Discuss the implications of such an action including at a minimum the impact on population growth, the economy, and US culture.

## Exercises

*Exercise 9.1*

For this assignment students as a team will examine the major demographic processes as they relate to the two counties chosen for this exercise. The objective is to develop an understanding of the ways in which demographic processes affect the populations of the counties under study. The final product should include at least 15 pages of narrative, along with relevant tables, graphs and maps.

The following processes should be addressed:

Migration:

- Identify current migration trends for the respective counties
- Compare the migration patterns for the two counties
- Identify historical trends related to migration for each of the counties
- Draw conclusions concerning migration patterns/trends for the counties under study.

## Reference

US Department of Homeland Security. (2022). *2021 statistical yearbook of immigration statistics*. https://www.dhs.gov/sites/default/files/2023-03/2022_1114_plcy_yearbook_immigration_statistics_fy2021_v2_1.pdf

## Additional Resources

US Census Bureau. *Current population reports.*

US Internal Revenue Service, County-to-County Migration. http://www.irs.gov/tax stats/article/0,,id=212695,00.html

Yusuf, F., Martins, J., & Swanson, D. (2014). *Methods of demographic analysis.* Springer.

CHAPTER 10

# Population Change

**Abstract** Population change is of particular interest to demographers. This chapter describes the process of population change and illustrates how the interplay of fertility, mortality and migration come together to determine its extent and direction. The distinctions between estimates, projections and forecast are discussed. The various ways of calculating population change are described and examples provided of various calculation methods. Trends in population change in the US and the world are described.

## Introduction

Population change is an inevitable aspect of social existence. Even the most stable traditional society will undergo change as a result of demographic processes, albeit sometimes at an imperceptible rate. Even in the absence of migration, differential birth rates or deaths rates for subgroups in society will result in compositional change. This is occurring in the United States today, for example, as the white population displays lower birth rates than various minority populations, leading to a restructuring of the racial and ethnic composition of the country's population.

Demographers pay considerable attention to population change since this phenomenon has implications for population size, population distribution, and population composition. A population's characteristics at any point in time represent a static "snapshot" of that population, and it is the changes that occur in it over time that represent the dynamic aspects.



R. K. Thomas, *Demography: An Introduction to Population Studies*, Springer Texts in Social Sciences, https://doi.org/10.1007/978-3-031-56623-3_10

There are three components to population change to consider: fertility, mortality and migration. Individuals are added to the population through the fertility process (births) and subtracted through the mortality process (deaths). Individuals are added to a population through in-migration and subtracted through out-migration. Population change resulting from the addition of births and the subtraction of deaths is referred to as "natural increase" (or decrease as the case may be). In a closed society, these are the only factors that influence population change. Since there are virtually no closed societies, the role of migration must be factored into the equation.

***There are three components to population change to consider: fertility, mortality and migration.***

## Theories of Population Change

Since demography is considered an applied science, little of this book is devoted to demographic theory. Nevertheless, there are certain theories that students of demographic should be aware of and some are addressed here. There are three generally accepted theories of population change: (1) Malthusian Theory; (2) Zero Population Growth; and (3) Cornucopian Theory; and 4) Demographic Transition Theory (OpenStax, 2021). The first three are briefly described below and then particular attention given to the demographic transition (see Box 10.1) since it is the theory to which demographers have made the greatest contribution.

According to Malthusian theory, three factors determine the Earth's carrying capacity for the human population, or how many people can live in a given area considering the amount of available resources. Malthus identified these factors as war, famine, and disease (Malthus, 1798). He termed them "positive checks" because they increase mortality rates, thus keeping the population in check. They are countered by "preventive checks," that control the population reducing fertility rates (such as birth control and celibacy). Malthus contended that people could produce only so much food in a given year, yet the population was increasing at an exponential rate. Eventually, he thought people would run out of food and begin to starve. They would go to war over increasingly scarce resources and reduce the population to a manageable level, and then the cycle would begin anew.

Of course, this has not exactly happened. The human population has continued to grow long past Malthus's predictions. There are three reasons why social scientists believe we are continuing to expand the population of our planet. First, technological increases in food production have increased both the amount and quality of calories we can produce per person. Second, human ingenuity has developed new medicine to curtail death from disease. Finally, the development and widespread use of contraception and other forms of family planning have decreased the speed at which the population increases. Some still believe Malthus was correct and that ample resources to support the earth's population will soon run out.

A neo-Malthusian researcher named Paul Ehrlich brought Malthus's predictions into the twentieth century. According to Ehrlich, it is the environment, not specifically the food supply, that will play a crucial role in the continued health of the planet's population (Ehrlich, 1968). Ehrlich's ideas suggest that the human population is moving rapidly toward complete environmental collapse, as privileged people use up or pollute a number of environmental resources such as water and air. He advocated for a goal of zero population growth (ZPG), in which the number of people entering a population through birth or immigration is equal to the number of people leaving it via death or emigration. While support for this concept is mixed, it is still considered a potential solution for addressing global overpopulation.

Cornucopian theory scoffs at the idea of humans wiping themselves out; it asserts that human ingenuity can resolve any environmental or social issues that develop. As an example, it points to the issue of food supply. If we need more food, the theory contends, agricultural scientists will figure out how to grow it, as they have already been doing for centuries. After all, in this perspective, human ingenuity has been up to the task for thousands of years. and there is no reason for that pattern not to continue.

Demographers are more likely to subscribe to demographic transition theory than other explanations offered, especially since there is clear worldwide evidence for this theory. Box 10.1 describes this prospective in detail.

**Box 10.1: The Demographic Transition**

Demographers describe the "demographic transition" (DT) as a process involving a population's transition from high birth and death rates to low birth and death rates as a society evolves from a pre-industrial society into an industrialized society. Demographic transition theory is based on an interpretation of demographic history developed by the American demographer Thompson (1929) based on his examination of changes in birth and death rates in industrializing societies over the previous 200 years. This process is typically represented by the demographic transition model (DTM) illustrated below.

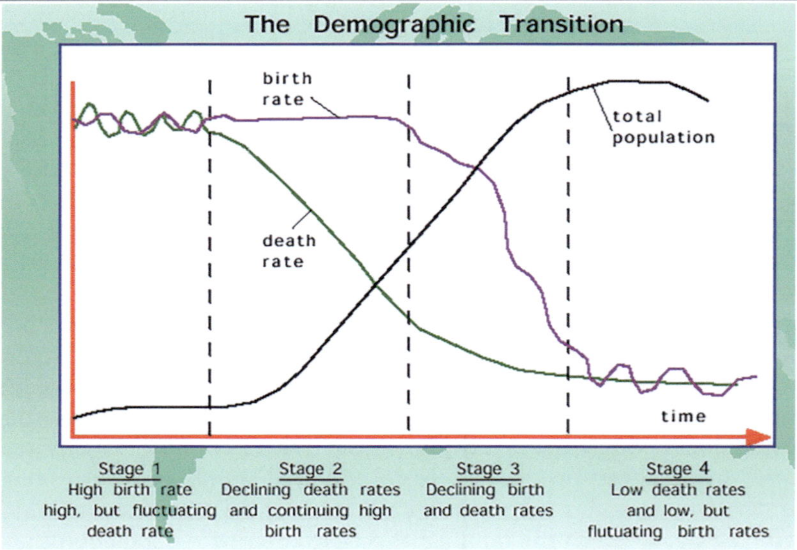


In Stage 1, for countries in the pre-industrial phase of development, deaths and birth rates are high and roughly in balance. All human populations are believed to have exhibited this balance of births and deaths before shifts began occurring in Western Europe in the late eighteenth century. Because both rates are approximately in balance, population growth is typically very slow in Stage 1.

In Stage 2 death rates drop rapidly as improvements in food supply and sanitation increase life expectancy and reduce the impact of disease. Without a corresponding fall in birth rates this produces an imbalance in vital events, and the countries in this stage experience a large and unprecedented increase in population.

In Stage 3, birth rates fall due to access to contraception, increases in wages, urbanization, a reduction in subsistence agriculture, and other social changes. Population growth begins to level off as a result. During Stage 4 both birth rates and death rates fall to unprecedented low levels. Birth rates, in fact, may drop to well below replacement level as has happened in countries like Germany, Italy, and Japan. Death rates may remain consistently low or increase slightly due to the emergence of lifestyle-related diseases and an aging population in developed countries.

By the late twentieth century, birth rates and death rates in developed countries had leveled off at rates that resulted in little or no population growth. Most developed countries today are in Stage 3 or 4 of the model; the majority of developing countries have reached Stage 2 or Stage 3. The only exceptions to this process are some poor countries in sub-Saharan Africa and the Middle East which remain in Stage 1.

As with all models, this is an idealized picture of population change. The model is a generalization that applies to groups of countries and may not accurately describe each individual case. The extent to which it applies to less-developed societies today remains to be seen. Many countries such China, Brazil and Thailand have passed through the Demographic Transition Model very quickly due to rapid social and economic change. Some countries, particularly African countries, appear to be stalled in the second stage due to stagnant development and the effect of AIDS.

Reference: Thompson (1929).

**Box 10.2: Demographic Processes: Interactions Among Fertility, Mortality and Migration**

Populations grow or decline, age or become younger and otherwise experience demographic change due to the interaction of the demographic processes of fertility, mortality and migration. Through most of human history, high fertility has combined with high mortality to keep population size and growth low. For example, between 10,000 BC and 5000 BC the population of the world grew from an estimated 1 million to 5 million (Vaughn's Summaries, 2011). At various points in time, mortality has spiked largely due to the spread of disease, climate change and/or war. High fertility could not offset the resulting large number of deaths and the overall population declined. In the eighteenth, nineteenth and twentieth centuries, high fertility combined with declining mortality to drive rapid population growth.

In all of these eras migration played an important role in the size and the growth of regional and local populations, most often as a response to changes in living and political conditions. Currently, low fertility is matched with low mortality to produce population loss in a number of nations in the developed world, while in the developing world fertility exceeds mortality by a relatively large margin. Migration continues to be a significant component of growth at the national and subnational levels in a number of nations.

Even at a time when the total number of humans alive on the planet was small, patterns of movement, for example, out of Africa to Asia and Europe, served to drive the growth of the population on the receiving end of the migration streams. Later on, migration streams arose out of Europe (voluntarily) and Africa (involuntarily), for example, to populate the Americas. From 1650 to 1820 the population of North America grew from about one-half million persons to 9.6 million. In some instances, the impact of migrants on the population at the receiving end was remarkable.

The combination of fertility, mortality and migration shapes not only the size of the population but its composition. Populations with high fertility and relatively low mortality such as those found in Mexico have what can be labeled as a young age structure. A young age structure and high fertility lead to significant population growth in both the short and long term. High fertility, low mortality populations tend to be young, and young populations produce large numbers of births in part because so many women are in their childbearing years. The social, economic and political issues that are linked to high and sustained population growth are complicated.

Although demographers have identified fertility, mortality and migration as key processes of population study, the important demographic outcomes of the interaction among those processes are the size, distribution and composition of populations whether they be for the world, nations, states within nations, or local areas.

While the measurement of population change is a critical activity of demographers, there are often situations in which adequate information is not available for this purpose. There are situations when a current population count is needed but not available due to the fact that no actual or estimated data can be obtained for the geography in question. There are other occasions in which information on population size and characteristics is desired for a past time period for which data are not available or for a future time period for which data are obviously not available.

In these situations, demographers rely on population estimates, projections and, to a lesser extent, forecasts. Using well-established techniques, data on population size and in some instances population characteristics are produced for past, present and future time periods. (Note that the range of techniques available for generating estimates, projections and forecasts cannot be fully covered in one chapter. The additional resources cited at the end of this chapter should be helpful in providing guidance to the wide range of techniques available for generating these statistics.) Box 10.2 discusses the interaction among the three processes that contribute to population change.

## Techniques for Calculating Population Change

### *Estimation Techniques*

Population estimates are typically calculated when there are no actual data available. Most often demographers speak of current estimates (i.e., for the current time period or one near the current time period). However, estimates may be made for past years when existing data points are available for past time periods.

***"Interpolation" and "extrapolation" are commonly used methods used by demographers to generate estimates and projections.***

Population estimates can be either simple or complex. The simplest approach to estimating or projecting change is to use known data points to extrapolate or interpolate data. Obviously, the more data points one has access to, the more accurate the estimate or projection. If one knows, for example the population of a community for 1990 and 2000, respectively, the population estimate for 1995 can be made by interpolating from these two figures. Box 10.3 discusses the interpolation and extrapolation processes that are employed in the generation of estimates and projections.

**Box 10.3: Using Interpolation and Extrapolation Techniques**
Demographers use interpolation and extrapolation as methods for generating estimates and projections. Most estimates employ an interpolation approach when existing data points can be accessed. This method assumes that a population figure for an intermediate year will fall somewhere between the populations identified for the beginning and ending periods. For example, if one were interested in a mid-decade population estimate (e.g., 2005), the estimate could be interpolated from the figures for 2000 and 2010. Theoretically, the mid-decade estimate would represent the mid-way point between the two figures (or, in this case, the average of the two). The mid-year population could fall outside the range of the two data points but this is assumed to be very unlikely.

When interpolation is used assumptions must be made, of course, concerning the rate of change for that decade, with the most straightforward approach assuming that the rate of change is constant over the time period in question. An assumption of this type must be made whether the rate of change is calculated in terms of numerical change or percentage change.

When calculating projections (and forecasts) the extrapolation method is typically used when past and/or current data are available. The rationale behind extrapolation is that, once a trend has been established, it can be expected to continue into the future on a similar trajectory. This is a "leap of faith" of course in that many factors can serve to divert a trend that has been underway for some time. Yet, the assumption of a continued trajectory is often made and, for the most part, can be empirically demonstrated. Thus, if an analysis wanted to project the population for 2020 for a defined population, it would be possible to extrapolate from existing data points into the future. The analyst could determine the numerical change between 2000 and 2010 based on census figures and assume the same absolute or percentage change for 2020. With this technique the 2000–2010 trend is extrapolated into the future. Interpolation could also be used for generating a projection if projections of populations for future years have been generated. For example, if the 2010 population count for the US is available and the Census Bureau has generated a projection for the US population in 2030, interpolation could be used to yield a projection for 2020.

Reality-Based Techniques. A number of methods are available for developing projections, with many of them categorized as "reality-based" techniques. These types of techniques tend to rely on "proxy" or "symptomatic" measures of population size and distribution. This approach utilizes data that are thought to be representative of the population based on some logical correlate. For example, some fairly direct measures of the number of households within a defined area might be determined by the number of utility hookups, on the assumption that all occupied dwellings require electricity, gas and/or water. Of course, assumptions have to be made with regard to household size since the only information available is the number of occupied units that are hooked up to the utility system.

A less direct method might involve the use of school enrolment data as a means of estimating population size. This method requires assumptions about the ratio of students to adults in a population and has to take into consideration private schools and home schooling. Data for proxy or symptomatic measures such as utility hookups and school enrolment are generally publicly available, providing an easily accessible source of proxy data.

***Reality-based estimates may use proxy or symptomatic data as a basis for generating population estimates in the absence of actual data.***

A reality-based method that is particularly useful in developing population estimates—assuming the requisite data are available—is the housing-unit method. This method has some similarities to the utility hookup estimates but is thought to provide a greater depth of understanding of the phenomenon under investigation. This approach involves starting with a known population—e.g., the "official" count from the most recent census—and employs an approach similar to the component method. Here, however, the "components" are housing units added to the community over a set period of time and housing units that are removed from the community over the same period. Thus, to the known number of housing units for Time 1 the analyst adds new home construction, new apartment construction, and newly established mobile homes between Time 1 and Time 2.

***There are often situations in which the desired data are not available, and demographers must generate "synthetic data" in the form of estimates, projections and forecasts.***

From the housing stock extant during Time 1 the analyst must subtract housing units (i.e., homes or apartments) that are demolished or otherwise taken out of stock and mobile homes that are eliminated. The required information on these activities is generally available from public records, although the availability, completeness and currency have to be considered on a case-by-case basis. The other factor that must be addressed, of course, is the average number of occupants for each type of housing unit. This information is generally available from American Community Survey data for the target area and can be applied to each type of unit. Finally, adjustments have to be made for occupancy rates for the various types of housing units since unoccupied housing will not contribute individuals to the population total. The figures generated for each type of housing unit are then summed to yield a population estimate for Time 2. While this approach is not able to take the homeless and migrant workers into consideration, for a "normal" population its specificity in terms of housing type makes it a useful technique.

Ratio Method. Another approach that is commonly used by demographers to generate population estimates and—less commonly—projections is the *ratio method*. This approach assumes that the share of the population for a geographical subunit represents a specified proportion of the total population. It further assumes that a constant ratio exists between the total population and the population for the sub-unit under study. Hence, past relationships between population growth in an area or community and that of its economic region or state are valuable guides for projection of population sub-units.

For example, for the United States, it is assumed that the population of a particular state represents a certain proportion of the total population and, further, that this ratio remains constant over time. Thus, assuming that one has an accurate count for the highest level of geography (e.g., the United States), it is possible to estimate the population for each of the states based on the population's ratio to the whole. If New York state historically accounts for 10% of the nation's population, it can be assumed, then, that it will account for 10% of the newly determined national population. The same procedure could be followed for any sub-unit, with county populations being derived from the state population, census tract populations being derived from the county population.

The ratio method for use in generating estimates can also be used for projecting populations for sub-units. The population growth of a study area can be projected into the future by relating its growth to a larger area of which it is a part, such as the state, the region, or the nation. The basic procedure is to compute the ratio between the population of the study area and some larger area for historical data points. This ratio may simply be between the study area and a larger area, or a series of interrelated ratios may be calculated between pairs of successively smaller geographical areas. Such a series, known as step-down ratios, might involve the study area and the State economic area, the State economic area and the whole State, the State and the region and, finally, the region and the Nation. This method requires the availability of a reliable projection for the larger area and comparable historical data for the subareas to be used.

## *Projection Techniques*

Projections are utilized to generate population statistics for future years. The data used for generating projections can be drawn from decennial census reports, the American Community Survey, and/or available local or state reports. Two data points (Time 1 and Time 2) are sufficient for generating a trend line and extrapolating into the future. If more data points are available all the better. In using the plotted information for projection purposes, the analyst assumes that the condition implied by the straight line will continue into the future.

Constant Change Method. The most parsimonious methods used for population projections are considered *constant change* methods of population extrapolation. These approaches assume that population growth follows natural laws and, therefore, can be expressed in mathematical or graphical form. Basically, the future population is forecast by examining past trends and projecting these trends into the future. The assumption is that the pattern observed for the past (e.g., increase, decrease) can be expected to continue into the future. This is typically a wholly mathematical exercise and does not consider the components that contribute to the observed change.

***In generating population projections demographers make assumptions about the extent to which observed historic trends can be expected to hold in the future.***

It is not necessary to be familiar with semi-log or other mathematical frameworks to generate straight-line projections and analysts may opt to use a method based on a constant arithmetic population change or a constant proportional population change. For the former, historic data which plot as a straight line on arithmetic graph paper imply constant arithmetic change for the population each year. This growth pattern implies that the population changes by the same number of people each year. The data would appear as a straight line when plotted on arithmetic paper. If it has been found, for example, that over a ten-year period the average annual increase was 1000 residents for the defined service area, then this figure (1000 per year) would be extrapolated into the future.

Not only does the approach assume that there is a constant average year-to-year increase but that the increase observed over the historical period was also constant year to year. For a defined geography it is found, for example, that the population increased by 10,000 over a ten-year period—or an average of 1000 per year. It is possible, however, that most of the increase occurred during the first years of the decade but slowed during the later years. Or, it could have been that the population was relatively stagnant during the first few years of the decade, with rapid growth occurring within the last few years. While the constant change method does not require that such factors be taken into consideration, if there is evidence of inconsistent change over the study period, the projection generated may be misleading.

An alternative to the numerical approach is the constant percentage change method. If one analyzed the numerical change results it would be found that, since the base population each year has increased (or decreased) by a constant amount, the rate of change is different for each year resulting in a different pattern of growth—that is, curvilinear). When the same data are plotted on semi-log paper (with population on the log scale and time on the arithmetic scale), a straight-line plots the results. In this situation, the numerical increase each year is greater than the year before, although the rate of increase is constant. Box 10.4 illustrates the results generated by using the two different models (numerical change and percentage change).

**Box 10.4: Constant Change Projection Models**

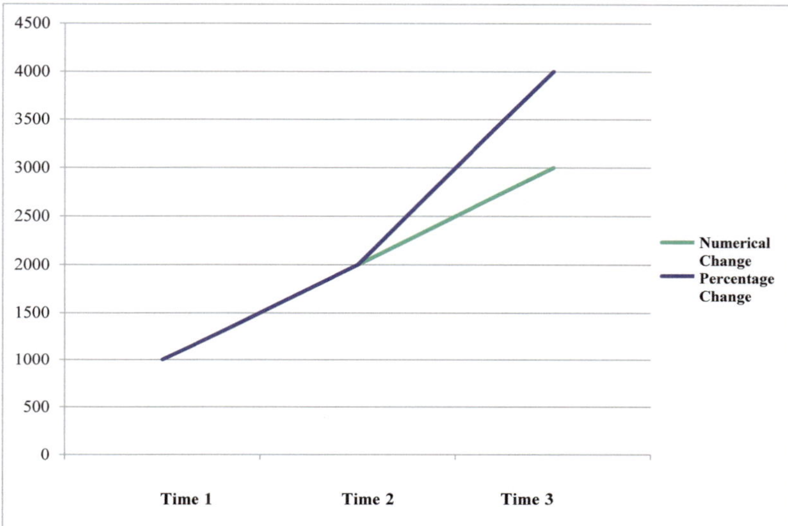


***The component method of calculating population change considers the three primary factors that influence population growth and decline—fertility, mortality and migration.***

Cohort-Component Method. One commonly used method for projections is the *cohort component method*. There are three components to population change to consider: fertility, mortality and migration. Individuals are added to the population through the fertility process (births) and subtracted through the mortality process (deaths). Box 10.5 provides formulas for the calculation of population change using the component method.

**Box 10.5: Calculating Population Change Component Method**

The component method for calculating population change uses births, deaths and migration flows as the basis for generating synthetic data. The first formula illustrates the calculation of natural increase/decrease while the second formula introduces migration into the equation.

Natural Increase/Decrease

$$Pop_2 = Pop_1 + Births - Deaths$$

Example:
$Pop_{2000} = Pop_{1990}$ + Births and - $Deaths_{1990\text{-}2000}$
??? = 10,000 + 6000 – 1000
$Pop_{2000}$ = 15,000
Natural increase = 5000

Natural Increase/Decrease and Migration

$Pop_2 = Pop_1$ + Births – Deaths + Migration
Example:
$Pop_{2000} = Pop_{1990}$ + Births and – $Deaths_{1990\text{-}2000}$
+ In-migration and – Out-Migration
$Pop_2 = Pop_1$ + Births – Deaths ± Migration
??? = 10,000 + 6000 – 1000 + 3000 – 1000 = 17,000
Pop $_{2000}$= 17,000
Population Change = + 7000

Cohort Change Ratio Method. Another available projection technique involves the use of *cohort-change ratios* (CCRs). This technique (also referred to as the Hamilton-Perry Model) can be used to develop projections not only by age, but also by age and sex, age and race, sex and race, and so on (Swanson & Tedrow, 2017). The major advantage of this method is that it has much smaller data requirements than the traditional cohort-component method. Instead of mortality, fertility, migration, and total population data, the Hamilton–Perry method simply requires data from the two most recent censuses. This method can be used to project population by age and sex using cohort-change ratios computed from data in the two most recent censuses or other data sets involving successive time periods.

In its simplest form this approach involves cohort-to-cohort advances over time. For example, the number of persons in the 20–24 age cohort in Time 2 should be similar to the number in the 10–14 age cohort tens year previous (if one is using decennial census data). The same is true for any other age cohort except, as noted below, the very youngest and very oldest cohorts. Assumptions must be made, of course, related to the impact of immigration and mortality on the population under study.

Given the nature of the CCRs, 10–14 is the youngest age group for which projections can be made (if there are 10 years between censuses). To project the population aged 0–4 and 5–9 one can calculate the Child Woman Ratio (CWR) and incorporate it into the model. This does not require any data beyond what is available in the decennial census. Projections of the oldest open-ended age group differ slightly from the CCR projections and require that the three oldest cohorts be grouped for calculating the population in the oldest cohort.

The cohort-change ratio from time period to period should equal 1.00 or less in the absence of any disruptive event. A cohort-change ratio greater than

1.00 indicates that more increments (e.g., in-migrants) then decrements (e.g., death and out-migration) occurred during the interval. Thus, for example, a fast-growing suburb would yield a higher ratio due to the fact that the numbers added to the population between censuses exceeded the numbers lost to death or out-migration.

One disadvantage of the cohort-change ratio method is that it can lead to unreasonably high projections in rapidly growing places and unreasonably low projections in places experiencing population losses. Geographic boundary changes are also an issue, even with census tracts. Since extrapolation methods of this type are based on population changes within a given area, it is essential to develop geographic boundaries that remain constant over time. For some sub-county areas, this presents a major challenge.

Predictive Modeling. Predictive modeling may be use to generate the number of people or events (e.g., disease incidence, crimes commited) anticipated for a future time period. The principal inputs for predictive modeling are: (1) National Center for Health Statistics (NCHS) rates for incidence/prevalence of disease and health services utilization (or other rates of occurrence); and (2) the population count for a specified population by age and sex in five-year intervals. Population counts may be keyed to a geographical area (including custom geographies) or any defined population. NCHS survey data provided through several recurring studies are utilized for this model.

The projection below involves the generation of data on the anticipated level of utilization of urgent care centers for selected diagnoses. This involves applying utilization rates for specific conditions to age/sex cohorts for census tracts.

The following steps are involved in applying the predictive model:

(1) Acquire population data for each of the nation's target geographies
(2) Identify the diagnosis codes that are relevant for this particular project
(3) Apply rates for these diagnoses for each age/sex category
(4) Generate output that displays (1) the number of anticipated cases for each diagnosis and (2) the rate per 1000 for each diagnosis.

Case Study 10.1 provides an example of the use of predictive modeling to anticipate the demand for health services.

**Case Study 10.1: Using Population Projections to Anticipate Health Services Demand**

Demographers involved in the planning of health services require access to data on disease incidence and prevalence, as well as data on the volume of health services consumed. However, there is no central repository of data on either the amount of morbidity within a population or the level of health services required to meet the needs of that population. Yet, this information is essential for the development of an informed health plan.

Health demographers do have some options due to the fact that the distribution of diseases in contemporary America exhibits a strong association with certain demographic characteristics. Unlike acute conditions, chronic conditions are highly correlated with the demographic attributes of a population. Thus, if the demographer has access to key demographic data for a defined population—e.g., the age and sex distribution—and utilization rates for various types of services it is possible to estimate and project health services demand for that population.

A case in point might be the need to estimate the demand for various physician specialties. A hospital, for example, may be initiating a physician recruitment effort or a chain of primary care clinics might be considering establishing a new site in a newly developed community. Assuming that the analyst has access to the age-sex breakdown for the target population it becomes possible to generate estimates and projections of the demand for various medical specialties among other factors.

Data are available from the National Center for Health Statistics that indicate physician utilization rates for each five-year cohort for males and females separately. It then becomes a simple matter of applying the age/sex-specific rates to each cohort to generate the anticipated number of cases. The estimates based of age and sex are typically adjusted for region of the country and, to the extent that additional attributes are available, they could be adjusted for race, income, insurance status or some other factor. Admittedly, these are modeled data but, in the absence of actual data, this represents the best possible estimate. Estimates and projections can be generated for any time period as long as the requisite inputs are available. The table below presents the anticipated number of visits for selected specialties for a specific metropolitan area for 2017.

**Estimated Physician Visits for Selected Specialties Metro Service Area 2017**

| | |
|---|---|
| Family Practice | 56,683 |
| Internal medicine | 28,450 |
| Pediatrics | 5157 |
| General surgery | 4257 |
| Obstetrics/gynecology | 12,347 |
| Orthopedic Surgery | 8447 |
| Cardiovascular Disease | 7687 |
| Dermatology | 7408 |
| Urology | 6043 |
| Psychiatry | 3322 |
| Neurology | 2917 |
| Ophthalmology | 19,583 |
| Otolaryngology | 3624 |
| All other specialties | 28,750 |

Not surprisingly, the calculations based on the available data suggest that primary care physicians (e.g., family practitioners, internists, OB/GYNs) are more in demand by this population than many of the specialists (e.g., psychiatry, neurology, otolaryngology). These estimates, in fact, reflect an older population that includes a lower than average proportion of children (hence fewer pediatric visits) and a higher than average proportion of seniors (hence, relatively more orthopedic, cardiovascular, and ophthalmological visits). Generating estimates in this manner allows the health demographer to use modeled data to fill a void in the existing knowledge.

### *Generating Forecasts*

*Forecasts* represent a special—and an infrequently used—form of synthetic data. Given the variety of factors that could contribute to population change, forecasts may make sense in some cases. Econometric approaches to forecasting use equations that project utilization as a function of the interplay of independent variables based on multiple factors. Theoretically, the more factors used in predicting future conditions, the more accurate the prediction will be.

Econometric Prediction. Econometric prediction incorporates anticipated contributors to population change in a series of mathematical expressions. The equation ultimately used is the one that best "fits the curve" exhibited by the historic trends. As a simple example, the historic ratios described above can be plotted as a time series and projected forward to generate a forecast. Local conditions must be examined, however, and the factors that might influence the future ratios fully understood. Former relationships between population growth in the area under consideration may suddenly change. Moreover, economic and social forces exert differing effects at different times on particular areas. Simply because a ratio has had a particular trend in the past is no assurance that it will continue to have that relationship in the future. For example, during the early decades of this century, coal mining towns in the Appalachian area grew at a faster rate than their State as a whole. However, during the past few decades, this trend has been reversed.

***Forecasts represent the most complex form of projections since they attempt to consider all possible factors that might influence population growth and change.***

Structural Equation Modeling. An example of a technique that could be used for demographic forecasting is structural equation modeling (SEM). The two central components of SEM are the path model and the measurement model. The path model or path analysis quantifies specific cause-and-effect relationships between observed variables. The measurement model quantifies linkages between (1) hypothetical constructs that might be known but unobservable components of the phenomenon under study and (2) observed variables that represent a specific hypothetical construct in the form of a linear combination. Cause-and-effect relationships between observed variables are usually based on theoretical considerations or evidence from prior studies. However, certain conditions must be met for a variable to be designated as cause versus effect.

Once the model is fitted to data, the path coefficient (direct effect) of each path is estimated and interpreted similarly to a regression coefficient. An indirect effect of any causal variable is estimated as the product of the chain of direct effects, and a total effect is the sum of all direct and indirect effects. As with multiple regression, an effect is interpreted as the change induced by fixing other variables in a model and changing only the subject variable. A direct effect would occur if all other variables in a model remained constant. In estimating an indirect effect, all other variables in the model are controlled except for the mediating variables in the path representing the indirect effect of interest. Observed variables should be both valid and reliable with respect to measurement of the latent variables.

The relationships in structural equation modeling are usually formulated by linear regression equations and can be graphically expressed by path diagrams. The distinct advantages of this technique compared to other statistical methods are the ability to test construct level hypotheses and provide information based on the analysis. As an example, this method could be used to explain the factors that determine a health index for an individual or, more appropriately, a population. A health index is often used as an indicator of the level of health of a community or a country. Besides being often used as an indicator of the state of welfare of a country, the health index may also indicate the level of productivity and economic growth of a particular country. If the relationship of the indicator variables with their respective latent variables could be determined the contributors to the health index could be identified.

This appears to be a particularly useful technique for forecasting a demographic-related phenomenon. The health of a population is influenced by factors that are related to the economy and the environment, as well as social and biological factors. Some of these factors are observable and some are not. The factors that are not directly observed, such as lifestyle, socio-demographic attributes, and mental health condition can be measured through indicator variables. Studies using this approach (e.g., Ferra et al., 2013) have found that structural equation modeling is a useful technique for generating demographic forecasts.

## The Impact of Population Change

Changes in population size almost invariably result in changes in population distribution and composition. Differential birth and death rates may cause changes in both distribution and composition. Similarly, the distribution of in-migrants may exhibit different patterns from the population in the receiving community. Migrants, in fact, affect the population composition of both the communities from which they originate and those to which they migrate.

In the United States historical migration patterns have contributed significantly to the redistribution of the US population. Migration from the eastern portions of the nation to the western portions has been a constant phenomenon since the continent's first settlers. More recently, migration from the Northeast and Midwest to the South and West regions has dramatically reshuffled the distribution of the nation's population and reshaped its social tapestry. Other examples of the impact of migration on the US population include the migration streams of primarily African-Americans from the South to the Midwest and Northeast during the period of agricultural mechanization during the first half of the twentieth century and the dispersion of the Hispanic population throughout the US since the 1990s. While Hispanics (particularly Mexican-Americans) remain concentrated in the West and southwestern states, migration over the past two decades has resulted in the spread of this population across the nation to the point that virtually every US County counts Mexican immigrants among its residents.

***A change in population size often involves changes in population distribution and composition.***

Change in population size typically involves changes in the composition of the population affected by the change. There are a few situations in which the population composition remains constant over time even as residential turnover occurs. Examples include a single-family neighborhood with facilities geared toward young families in which the residents' characteristics remain constant over time as new young families replace older families whose housing needs have changed. Or consider the retirement community that has age restrictions and perhaps other requirements of its residents to assure a consistent demographic mix for the community.

In most cases, however, it is expected that population turnover will be accompanied by changes in population composition. This has been demonstrated time after time in our urban centers as wave after wave of successive immigration groups have one after another displaced each other, thereby significantly changing the character of the community…or as "urban pioneers" replace long-time residents in inner cities through the gentrification process. An important role for applied demographers involves the ability to anticipate the implications of population change, and Case Study 10.2 describes the impact of residential turnover on population composition.

**Case Study 10.2: Population Turnover and Compositional Change**

When population change occurs it not only affects the size of the population but also has implications for population distribution and composition. Whether the change is engendered by persons added to the population through birth or in-migration or persons subtracted from the population through death or out-migration, population size, distribution and composition are inevitably affected.

A case in point is an older suburban area on the fringe of the inner city in a medium-sized Southern city. The Taylor community had been established during World War II as an industrial community of middle- and working-class families. It maintained this status until the 1970s when the last plants closed and jobs disappeared from the community. The decline of industry coincided with the beginning of the movement of inner-city residents into near suburbs and the subsequent flight of white residents to more distant residential areas.

The residential turnover that occurred between 1980 and 2010 had a major impact on the composition of the residents of the Taylor community. In terms of population size, there was little change over the two decades with the number of people residing in the community in 2010 little different from that in 1990. The distribution of the population geographically within the community changed little over these two decades as well, since the area was essentially "built out" and there was little space for additional residential development.

A major shift, however, was noted in the composition of the population. By the time the influx of inner-city residents was underway, the original Taylor population had aged considerably, with many of the original "settlers" from the 1950s and 60s aging in place. Older residents made up the bulk of long-term community members while the younger age cohorts (e.g., ages 5–34) were overrepresented among the newcomers. Interestingly, the sex ratio exhibited little change over this two-decade period.

The most important change was recorded for the racial mix. In 1990 the white population still maintained a slight edge (55–45%), but by 2010 the population was 89% black and 11% white. No other racial or ethnic group was represented to any extent. This racial turnover was accompanied by a number of other changes. The proportion of the population that was married dropped from 55 to 32% (with one-fourth of the latter being separated). The proportion never married increased from 28 to 51%. The proportion of family households declined from 79 to 71% and the proportion of female-headed households with children increased from 17 to 26%.

Similar dramatic changes were recorded over the two decades for a number of other attributes. The median household income actually declined between 1990 and 2010 (from $23,270 to $21,367) as did the median family income (from $27,942 to $24,750). At the same time, the poverty population increased dramatically, with the proportion of persons living at or below poverty increasing from 20 to 61%, the proportion of poverty-level families from 17 to 36% and the proportion of impoverished children under 18 from 21 to 63%.

These changes in socioeconomic status reflected a decline in labor force participation overall (from 67 to 62%) but an increase in female labor force participation (from 24 to 44%). The already high unemployment rate increased dramatically (from 15 to 40%). The educational level did not increase over this time period and actually declined slightly.

Clearly, as a result of population change, the Taylor community looked a lot different in 2010 than it did in 1990. It was transformed from an older, predominantly white moderately educated middle- and working-class community of stable families into a predominantly younger African-American lower- to working-class population with limited education and high rates of unemployment. While the overall size of the population and its distribution felt limited effect from the population turnover that occurred, the population composition looked very different in 2010 than it did in 1990.

## Data Sources for Population Change

Demographers often rely on government sources of population data, and the federal government (by virtue of the Census Bureau) is the major source of demographic data. However, the now limited data generated through the decennial census are not available for a year or often longer after its collection and are quickly dated. Data from the American Community Survey are also not available for a year or more after collection. While the Census Bureau and state data agencies may calculate projections for future years they seldom provide data below the county level. For data at the sub-county level non-governmental sources may have to be accessed. For these reasons, demographers must acquire the skills necessary to produce estimates and projections.

***The Census Bureau is the major source of data on population change.***

Population estimates for states, MSAs, and counties are prepared each year as a joint effort of the Census Bureau and the state agency designated by each state governor under the Federal-State Program for Local Population Estimates (FSCPE). The purpose of the program is to standardize data and procedures so that the best quality estimates can be derived. Most states also generate population estimates and projections that are available through state agencies. However, these figures are often produced at irregular intervals, and thus may be quite dated. The reader is also encouraged to evaluate the quality of these data to the best of his or her ability.

Population estimates and projections generated by government agencies have historically been the only ones available. Today, however, a number of data vendors provide these figures. These vendor-generated data are often made available down to small units of geography (e.g., the census tract) and in greater detail (e.g., sex and age breakdowns) than government-produced figures. They offer the flexibility to generate estimates and projections for "custom" geographies (e.g., a market area) not available for government-generated statistics.

The drawback, of course, is that some precision is lost as one develops calculations for lower levels of geography and for population components. However, the ease of accessibility and timeliness of these vendor-generated figures have made them a mainstay of those conducting demographic analyses.

Statistics on births and deaths are readily available, as are statistics on international migration. The real challenge with regard to internal migration is finding the people to count. While immigrants are tracked fairly accurately, there is no similar mechanism for identifying and counting internal migrants. This information has to be gathered primarily through surveys. The only near-comprehensive source of data on internal migration is the year-to-year tracking of county-to-county movement on the part of the Internal Revenue Service. Obviously, those who do not file taxes are excluded, and the IRS only published data on origin county, destination county, household size, and household income.

***A number of data vendors provide projections, with these figures often made available down to small units of geography.***

Issues have been raised concerning the quality of the synthetic data produced by both government agencies and commercial data vendors. Data users typically need the latest information possible, and in an effort to be expedient the question of quality sometimes has become a secondary concern. Any evaluation of the quality of synthetic data requires an understanding of the recency and quality of the historical data being used as a basis for the estimates and projections. Furthermore, attention must be paid to the methods and assumptions utilized to generate the figures.

If, for example, one assumes that population growth in an area is gradual and can be described by a simple mathematical function, population estimates and projections will be reasonably accurate as long as the assumptions hold. However, to the extent that an assumption is wrong, the (incorrect) mathematical function will yield inaccurate estimates and projections. While it is not possible to be aware of all the nuances of data quality and method, users are urged to evaluate underlying assumptions critically and to ascertain the accuracy of the synthetic data that are available.

## Key Points

- There are three components of population change to consider: fertility, mortality and migration.
- These three components interface in complex ways in order to generate population change.

- Countries experience the demographic transition as they move from high birth rate/high death rate circumstances to low birth rate/low death rate circumstances.
- Estimates, projections and forecasts are referred to as synthetic data since they generate data in the absence of actual numbers.
- Estimation techniques are used to generate estimates for past time periods or for the current period and rely heavily on interpolation methods.
- Projection techniques are used to generate data on demographic attributes in the future, often employing extrapolation methods.
- Reality-based estimates may use proxy or symptomatic data as a basis for generating population estimates.
- The component method of calculating population change takes into account all three processes involved in population change.
- The cohort-change method involves a straightforward technique for estimating population change.
- Predictive modeling is a method for applying projection techniques to real-world problems.
- Forecasts represent the most complex form of projections since they consider all factors that might influence the nature of change.
- A change in population size often involves changes in population distribution and composition.

## Discussion Questions

10.1 Students should review the two different types of constant change projection techniques. Consider the pros and cons of each and their advantages and disadvantages vis-à-vis other more complicated methods. Which is the best constant change method for most purposes?

10.2 Considering the types of synthetic data discussed here, what are the circumstances under which one would generate an estimate rather than a projection rather than a forecast. Present examples of each of the different approaches.

## Exercises

*Exercise 10.1: Constant Change Projections*

There are two ways in which a constant change technique can be used to generate population projections: numerical change and percentage change. For the community featured below examine the data provided and compute projections using these two different techniques. Discuss how the outcome differs depending on the technique used.

Model 1: Numerical change

Population $T_{1990}$ = 10,000   Population $T_{2000}$ = 20,000   Population $T_{2010}$ = ________

Calculate the average annual numerical change between 1990 and 2000. Apply this figure to the period between 2000 and 2010 and calculate the 2010 population.

Model 2: Percentage change

Population $T_{1990}$ = 10,000   Population $T_{2000}$ = 20,000   Population $T_{2010}$ = ________

Calculate the average annual percentage change between 1990 and 2000. Apply this figure to the period between 2000 and 2010 and calculate the 2010 population.

Examine the two outputs and determine how they differ. Plot the trends using an Excel spreadsheet or other tool and compare the slopes generated by the two techniques.

*Exercise 10.2: Calculating Population Change*

You have been asked to calculate population change for Podunk County in order to develop plans for school expansion. Calculate population change first by looking only at natural increase and, then, by including migration in the equation.

Calculate the missing values for Podunk County based on natural increase for a future year and for a past year:

| *2005 Population* | *Births 2005–2014* | *Deaths 2005–2014* | *2015 Population* |
|---|---|---|---|
| 50,000 | 500 | 200 | ______ |

| *2005 Population* | *Births 2005–2014* | *Deaths 2005–2014* | *2015 Population* |
|---|---|---|---|
| ______ | 500 | 200 | 30,000 |

Calculate the missing values for Podunk County based on natural increase AND migration:

| *2005 Population* | *Births 2005–2014* | *Deaths 2005–2014* | *In-migration* | *Out-migration* | *2015 Population* |
|---|---|---|---|---|---|
| 50,000 | 500 | 200 | 1000 | 200 | ______ |

| *2005 Population* | *Births 2005–2014* | *Deaths 2005–2014* | *In-migration* | *Out-migration* | *2015 Population* |
|---|---|---|---|---|---|
| 50,000 | 500 | 200 | 200 | 800 | _______ |

| *2000 Population* | *Births 2005–2014* | *Deaths 2005–2014* | *Net migration** | *2015 Population* |
|---|---|---|---|---|
| 50,000 | 500 | 200 | _____ | 60,000 |

*Net migration = Difference between in-migration and out-migration

*Exercise 10.3: Housing Unit-Based Estimates*

Students are tasked with the job of estimating the population for Smith County for 2015 using a reality-based approach. Information is available on the number of housing units of various types in the county in 2010 based on the decennial census. Information is available from local sources on the new housing units that have been added since 2010. Estimates of the average household size can be obtained from the American Community Survey (ACS). Using the information provided, generate an estimate of the 2015 population of Smith County.

Smith County 2010 population = 20,000.

Housing units added between 2010 and 2015 and the average household size:

| *Housing type* | *Number* | *Average Household size* | *Added Population* |
|---|---|---|---|
| Single-family home | 1000 | 3.00 | ______ |
| Apartments | 500 | 2.00 | ______ |
| Mobile homes | 50 | 2.50 | ______ |
| Total population added between 2010 and 2015 | | | ______ |
| Smith County 2015 Population = 20,000 + [added population] = __________ | | | |

*Note* This exercise does not take into consideration housing stock that has been removed from the total between 2010 and 2015.

## References


Ehrlich, P. R. (1968). The Population Bomb. New York: Ballantine Books.

Ferra, Z., Kamarulzaman, I., & Jemian, A. A. (2013). *Bayesian structural equation modeling for the health index*. https://ukm.pure.elsevier.com/en/publications/bayesian-structural-equation-modeling-for-the-health-index

Malthus, T. R. & Malthus (1798) An Essay on the Principle of Population, 1st ed. London: J. Johnson.

OpenStax. (2021). *Introduction to sociology*. https://openstax.org/details/books/introduction-sociology-3e

Swanson, D. A., & Tedrow, L. (2017). *The top ten reasons to use the cohort change ratio method*. file:///C:/Users/Thomas%20pc/Desktop/SWANSON%20&%20TEDROW%20PAA%202016%20V1%20(2).pdf

Thompson, W. S. (1929). Population. *American Journal of Sociology., 34*(6), 959–975.

Vaughn's (2011). World Population Growth History Chart. Vaughn's Summaries. Downloaded from https://www.vaughns-1-pagers.com/history/world-population-growth.htm#population-growth-chart.

## Additional Resources

Baker, J., Swanson, D. A., Tayman, J., et al. (2017). *Cohort change ratios and their applications*. Springer.

Smith, S. K., Tayman, J., & Swanson, D. (2014). *A practitioner's guide to state and local population projections*. Springer.

CHAPTER 11

# United States Demographic Trends

**Abstract** This chapter expands on previous chapters as it describes the various demographic trends that are affecting US society. The US population in the 2020s is almost unrecognizable to those with a twentieth century perspective. Virtually every demographic trend has been affected by developments in US society. The population has changed to the point that there are an increasing number of "firsts" when it comes to demographic attributes. The impact of past trends is examined, and the anticipated implications of current trends explored.

## Introduction

The demographic trends that have been playing out over the past quarter of a century are dramatically reshaping the US population and the nature of American society. Even the trends themselves—particularly the key demographic processes—would be unrecognizable to our forefathers. Not only have these developments created a population with traits that are mostly unprecedented, but their unfolding has consequences for every American institution.

American society is no stranger to change and, indeed, change has become a value in its own right in contemporary America. Yet the developments occurring over the past 25 years are not extensions of previous trends but are unique in our history and unprecedented in their impact. These demographic trends are important in that they have contributed to the changing composition of our population; this, in turn, has influenced the processes that brought about the changes. These shifts also have implications for the future demographic profile of the American people. Indeed, the demographic transformation of the American population in the late twentieth and early twenty-first centuries



R. K. Thomas, *Demography: An Introduction to Population Studies*, Springer Texts in Social Sciences, https://doi.org/10.1007/978-3-031-56623-3_11

might be considered a major, if not *the* major, determinant of the challenges to be faced by our nation's institutions in the future.

***The demographic developments occurring over the past 25 years are not extensions of previous trends but are unique in our history and unprecedented in their impact.***

This chapter cannot begin to address all of the demographic trends that have contributed to the changing social environment. It emphasizes the key demographic trends and their implications for American society.

## Demographic Processes

The first step in understanding contemporary America is to examine the basic demographic processes as they have played out over the past quarter century. These are the trends that are driving population change and remaking our demographic profile. At some level we are aware of these trends, since they are mostly the results of our actions as members of society. We might, however, not be aware of the far-ranging implications of the developments that are discussed below.

### *Fertility*

Americans like to think of themselves as a young, virile population. Even today many advertisements present our increasingly mature population in the guise of young models. However, our marketing gurus are finally recognizing the fact that we are no longer a nation of youngsters. Seniors today are being swamped with solicitations for nursing homes, assistive equipment and erectile disfunction medications at a frequency unthinkable in the past.

Compared to our own historical rates and to those of much of the rest of the world, the level of US fertility is quite low today. In fact, we are not producing enough babies to replace those we are losing due to death. If it were not for immigration adding new bodies to our population, we would be exhibiting population decline like many developed countries. In order to replace ourselves, American women need to produce 2.1 births over the course of their lifetimes. We have not reached that level in years and, even when we were at replacement levels 20 years ago, much of that reflected the last surge of baby boom women in their 30s and 40s having babies, coupled with an increase in the number of births to immigrant women.

***Compared to our own historical rates and to those of much of the rest of the world, the level of US fertility is quite low today.***

The total number of births increased from 2.9 million in 1945 to 4.3 million in 1960, a figure that has only been approached one time since. The period from

1946 to 1964 is generally recognized as the era of the post-World War II *baby boom*; until 1989, 1964 was the last year in which at least four million births were recorded for American women. The interval between 1965 and 1972 is seen as the transition to the *baby bust* that lasted from 1972 to 1978. Although some regard post-1978 fertility as evidence of a baby boomlet, it might be better described as an *echo baby boom*. In other words, there was an increase in births due to the rapid rise in the number of potential and, then actual, mothers as the children of early baby boomers reached their childbearing years. This, however, reflects an increase in the number of potential mothers rather than a rise in fertility rates.

After 1987 there was an increase in the number of births that could not be explained by the echo effect. This was partially driven by baby boom women who were having children at older ages than previous generations of mothers. By 2007, the figure for annual births had reached 4.3 million, a number not realized since the 1957–1961 period. This surge in births created what has been described as the *millennial generation* (or *Generation Y*), with its 80 million members representing the largest cohort since the baby boomers. This cohort born between 1981 and 1997 was succeeded by a smaller cohort as the number of births began falling (despite continued population growth).

Over a five-decade span, the United States experienced a peak of 4.3 million annual births (in the early 1960s), a valley of 3.1 million births (in the mid-1970s), a rise to about 4.0 million births (during the early 1990s), a small decline to 3.9 million (in the late 1990s), another peak to 4.3 million in the early twenty-first century, and a significant drop in the wake of the 2008 recession. Even with a growing population the number of births had dropped back down to 3.6 million by 2020. Box 11.1 presents trends in births and birth rates for the US from 2000 to 2021.

***While Americans have been recording relatively high numbers of births, they are also recording record low fertility rates.***

While the total number of births in the US today is approximately the same as it was at the peak of the baby boom, the average number of children women bear over a lifetime is now about 60% of that of the early 1960s; in 2020 the average woman was estimated to bear 1.6 children over her lifetime. While Americans were recording relatively high numbers of births, they were also recording record low fertility rates.

**Box 11. 1: Number of Births and General Fertility Rate United States 2000–2021**

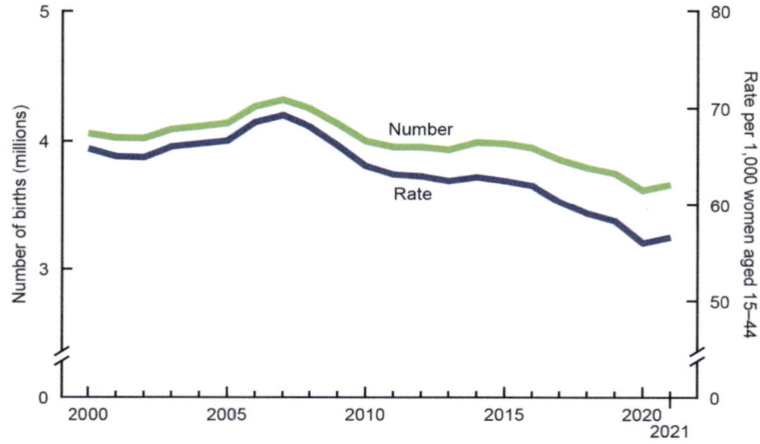


*Source* National Center for Health Statistics, National Vital Statistics System, Natality

The total fertility rates (TFR) for the African American and Hispanic populations are somewhat higher than that for whites, and, as the US population becomes more heavily populated by these two minority groups, the TFR may rise without any real change in fertility. In 2020 the anticipated TFR for non-Hispanic whites was 1.6, compared to 1.7 for African Americans, 1.4 for Asian-Americans, and 1.9 for Hispanics. The American (fertility) dream is alive and well but being realized by new and different segments of the population. Hispanics and African Americans are outproducing the non-Hispanic white population, although even the rates for these segments of the population are exhibiting decline. Box 11.2 presents birth statistics for key racial and ethnic groups. It is noteworthy that all of these figures have declined since the beginning of the twenty-first century.

***Higher total fertility for Hispanics and African Americans Means they are out-producing the non-Hispanic white population.***

There are several trends related to birth outcomes that are important to consider. Not only has contraceptive use increased since the 1970s, but the pattern of use has changed over time. In 2019, 65% of women of childbearing age used some type of nonsurgical contraceptive (e.g., oral contraception, intrauterine device [IUD] or diaphragm). As of that date, 18% had been sterilized (Daniels & Abma, 2020). Reliance on the pill, IUD, and diaphragm has declined since the 1970s, while sterilization as a means of contraception has become increasingly common. There remain a large number of women who do not use contraceptives, leaving them at risk for both wanted and unwanted pregnancies. Nevertheless, the widespread use of contraceptives has contributed to the decline in birth rates.

**Box 11.2: Birth Rates for Selected Racial and Ethnic Groups United States: 2020**

| | *Crude Birth rate* | *General Fertility rate*** | *Total Fertility rate*** |
|---|---|---|---|
| Total | 11.0 | 55.8 | 1641 |
| Non-Hispanic Whites | 9.5 | 53.2 | 1552 |
| Hispanics | 14.1 | 62.8 | 1876 |
| Blacks | 12.9 | 59.0 | 1714 |
| Asian-Americans | 11.6 | 50.1 | 1385 |

*Per 1000 population
**Per 1000 women 15–44
*Source* National Center for Health Statistics

A change in the average age at first intercourse can have important implications for fertility patterns. The rate of premarital intercourse on the part of teenagers has increased markedly since the 1970s. Earlier age at first intercourse results in an increase in the risk of pregnancy and concomitant increases in health risks. Despite the increasing younger age of first intercourse, the birth rate for women 15–19 has steadily declined largely due to increases in the proportion using contraceptives.

Another major trend in fertility relates to children born to unmarried women. In 2020, nearly 41% of all births in the United States were to unmarried mothers, representing a dramatic increase over the decades following WWII. Approximately 70% of all African American births are to women who are unmarried, but only 28% of white births are so classified (National Center for Health Statistics, 2022). The figure for Asian-Americans was only 17%, while Hispanics recorded a higher-than-average rate of 53%. While the proportion of African-American births occurring out of wedlock appears to have leveled off (albeit at a very high rate), the proportion of white and Hispanic births to unmarried mothers continues to increase. Box 11.3 presents recent data on the percentage of all births to mothers who are unmarried.

***In 2020, nearly 41 percent of all births in the United States were to unmarried mothers, representing a dramatic increase over previous generations.***

The long-term fertility pattern for the US reflects a major reversal of historic patterns and raises the question of how the US moved within two decades from a fertility rate considered high by the standards of industrial nations to one of the lowest. From the late 1940s to the early mid-1960s we reproduced at record rates. Part of the surge, of course, was the pent-up demand after World War II. This coupled with unprecedented economic prosperity and a new future-oriented mindset produced the baby boom. Then, a number of trends acting in concert put a damper on American reproductive activity.

**Box 11.3: Births to Unmarried Women United States 1980–2021**

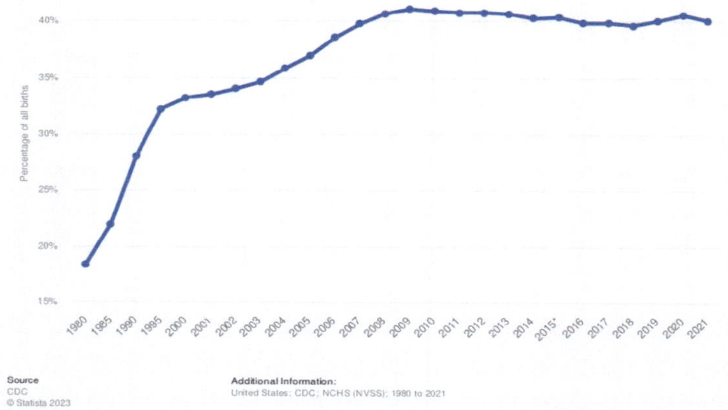


For one thing, people's lives became much more complicated. The idealized life of the 1950s depicted on television (working father, homemaking mother and 2.3 children) seems like a mirage today and (according to demographers) probably is. Women entered the workforce in unprecedented numbers beginning in the 1960s. This was not just single women who needed to make a living either, it increasingly included married women *with children*—something pretty much unheard of a few years earlier. (Unheard of, that is, among middle-class white citizens. It was probably always the case with minority mothers.)

***The long-term fertility pattern for the US reflects a major reversal of historical patterns.***

For this and other reasons, men and women began to put off marriage to a much later age than during the post-war period. Part of this, of course, was a function of increased college enrollment. Women in particular who had been marrying (and having children) right after high school now found a reason to delay both marriage and childbearing. In 2021 the average age of first marriage in the United States was 28.6 years for women and 30.6 years for men, up from 20 and 22 in 1960.

***Women entered the workforce in unprecedented numbers beginning in the 1960s, thereby making a major contribution to declining fertility rates.***

Among women who married, two other factors have contributed to the fertility downturn. Couples were deciding to have fewer children—or none at all. By 2000 the most common family type was married couples *without children*. The other factor, which partly contributes to the first, is the decision among those who did desire children to delay childbearing until later years.

One other consideration when it comes to reproduction was the significance of abortion. The number of legal abortions increased steadily during the 1970s and 1980s and leveled off thereafter. A large number of abortions, it could be argued, suppresses the number of births. If one examines trends in abortions—noting that records are kept only on legal abortions—it is found that in many years more than one million births were precluded due to the use of abortions as a means of termination. Now, it is likely that some portion of the aborted fetuses would not have survived to term (that is, been miscarried) or not have survived the first year of life. Still, during the peak year for abortions over 1.5 million potential births were prevented. Box 11.4 indicates trends in legal abortions in the US for 1973–2020.

**Box 11.4: Trends in Abortions United States 1973–2020**

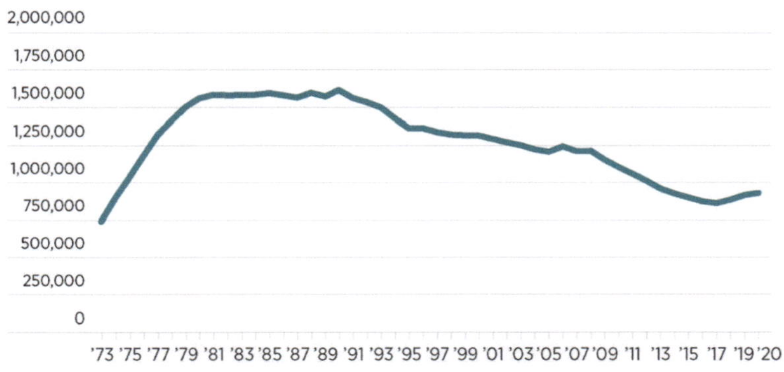


*Source* Guttmacher Institute

The number of legal abortions climbed steadily from 200,000 in 1970 after the liberalization of abortion laws to a high of around 1.6 million in 1990. Since 1990 the number has steadily decreased, representing a decline of about 400,000 annually from the 1980s. The number of abortions recorded rose unexpectedly in 2020. The 2023 ruling by the US Supreme Court negating longstanding federal protection for abortion is expected to have a significant—although as yet unknown—impact on the number of abortions performed.

### *Mortality*

Like fertility, mortality is a dynamic process, influenced by a number of demographic factors and subject to significant fluctuation. The total number of annual deaths in the United States increased steadily during the twentieth century. More significant than the number of deaths, however, is the mortality rate since it is adjusted for population size. The crude death rate in 1900 of 17.2 per 1000 declined dramatically during the first half of the twentieth century, with a rate of 10.8 recorded in 1940. The death rate continued to decline during that century, dropping to around 8.7 per 1000 by the end. However, there has been a recent increase in the death rate, with the most recent NCHS data indicating a 2021 rate of 10.5 per 1000.

Ironically, the age-adjusted death rate followed a different course, rising to 8.8 per 1000 in 2021. The figure remained higher for males than for females. African Americans continue to report the highest age-adjusted death rates, while non-Hispanic whites report slightly higher than average rates. The age-adjusted death rates for Hispanics and Asian-Americans are both lower than the rate for non-Hispanic whites. These rates are presented in Box 11.5.

**Box 11.5: Age-Adjusted Death Rate By Race and Ethnicity United States 2020**

| *Group* | *Deaths rate*[a] |
|---|---|
| Non-Hispanic White | 8.4 |
| African American | 11.0 |
| Asian/Pacific Islander | 4.7 |
| American Indian/Alaskan Native | 10.1 |
| Hispanic | 7.2 |

[a]Deaths per 1000
*Sources* National Center for Health Statistics

Over the course of the past century, life expectancy steadily increased, rising from around 49 years at the beginning of the twentieth century to 78.5 years in 2000. In an unexpected development, life expectancy has actually declined for the US population since then, with a life expectancy of 77.3 recorded for 2020. This decline not only reflects the unhealthy lifestyles characterizing many Americans but indicates the impact of the COVID-19 pandemic and the increase in "deaths of despair". The figure for 2020 does not even reflect the full impact of the pandemic on life expectancy.

Asian-Americans now report the highest longevity among various racial and ethnic groups. Non-Hispanic whites historically had this distinction (with white females reporting the longest lives on average), but the advantage for whites is increasingly being diminished as other racial and ethnic groups report increased life expectancy. It should be noted for all groups—and especially non-Hispanic whites—the rate of increase has slowed significantly over past decades, suggesting that a limit has been reached in terms of improving longevity (at least for the US population). Life expectancy for females continues to be higher than that for males. Box 11.6 illustrates this trend graphically up through 2020.

**Box 11.6: Trend in Life Expectancy United States 1980–2020**

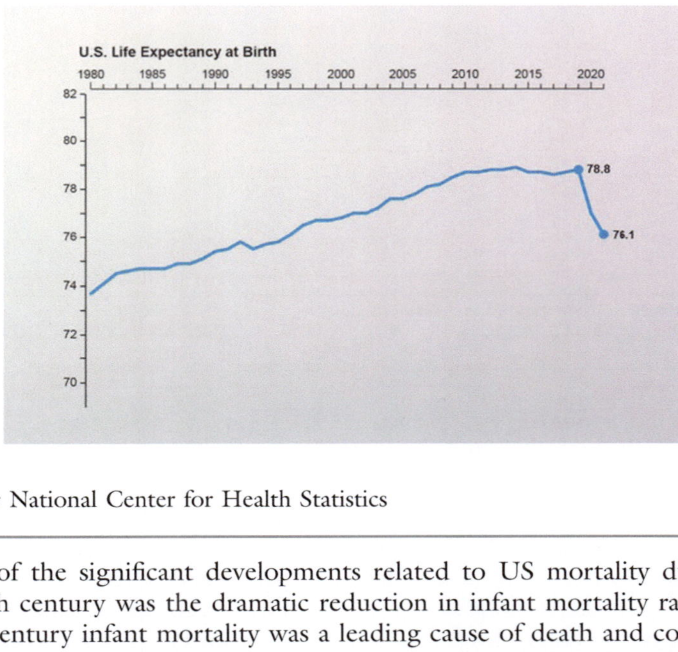


*Source* National Center for Health Statistics

One of the significant developments related to US mortality during the twentieth century was the dramatic reduction in infant mortality rates. Early in that century infant mortality was a leading cause of death and contributed to a relatively short average life expectancy. Between 1920 and 1940 the IMR was essentially cut in half and by 1970 it was halved again. Overall, a rate of 85 infant deaths per 1000 live births in 1920 was cut to 7 per 1000 by 2000 and to 5.4 by 2020.

A similar but even more dramatic trend was recorded for maternal deaths, a leading cause of death for females at the beginning of the twentieth century. The maternal death rate of around 80 per 10,000 live births was reduced to only 1 per 10,000 by the end of the twentieth century. The sharp reduction in the maternal mortality rate was directly responsible for the rapid increase in female life expectancy that took place in the middle part of the twentieth century. The improvements in infant and maternal mortality rates can be attributed to a number of factors in addition to better medical care. These include improved public health conditions and changes in demographic characteristics such as education and income. Widespread access to prenatal care has also played a role. Box 11.7 depicts the trend in US maternal mortality compared to comparable countries.

**Box 11.7: Maternal Mortality (per 100,000 live births) United States and Comparable Countries 1990–2017**

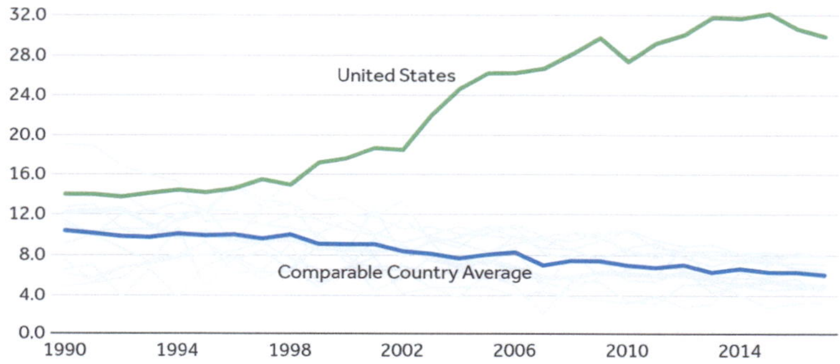


Note: IHME's methods of determining pregnancy-related deaths and the age ranges included differ from other estimates.

Source: Kaiser Family Foundation analysis of data from Institute of Health Metrics and Evaluation, Global Burden of Disease Study 2017 (GBD 2017) Data Downloads.

Peterson-KFF
**Health System Tracker**

***The maternal mortality rate for the US has reversed course and is increasing to a level not seen in decades.***

Past improvements in infant and maternal mortality have both been tarnished by recent developments. The decline in US infant mortality rates has leveled off, while those for other developed countries have continued to decline. Even more disturbingly, the maternal mortality rate for the US has reversed course and is increasing to a level not seen in decades. Although the number of birth-related deaths for mothers remains low, the fact that the US rate is rising when that for all other comparable counties is declining is reason for concern. Box 11.8 presents trends infant mortality and maternal mortality for the past 10 decades.

**Box 11.8: Infant and Maternal Mortality Rates in the United States 1920–2020**

| *Year* | *Infant mortality*[a] | *Maternal mortality*[b] |
|---|---|---|
| 2020 | 5.4 | 2.3 |
| 2015 | 5.9 | 1.8 |
| 2010 | 6.1 | 1.3 |
| 2005 | 6.9 | 1.0 |
| 2000 | 6.9 | 1.0 |
| 1990 | 9.2 | 0.8 |
| 1980 | 12.6 | 0.9 |
| 1970 | 20.0 | 2.2 |
| 1960 | 26.0 | 3.7 |
| 1950 | 29.2 | 8.3 |
| 1940 | 47.0 | 37.3 |
| 1930 | 64.6 | 67.3 |
| 1920 | 85.8 | 79.9 |

[a]Deaths per 1000 live births
[b]Deaths per 10,000 live births
*Source* National Center for Health Statistics

Significant changes have occurred with regard to the major causes of death affecting US society. The dominant causes of death early in the twentieth century have waned in importance due to improvements in disease prevention, including immunization, and treatment. Chronic diseases like heart disease and cancer now account for about 48% of all mortality. This trend (referred to as the "epidemiological transition) reflects a shift from a predominance of acute conditions to a predominance of chronic conditions. Box 11.9 compares death rates from various causes over time.

### *Significant changes have occurred with regard to the major causes of death affecting US society*

While acute conditions like the common cold come and go with great frequency and most people are affected at some time or another, chronic conditions like heart disease, diabetes, and emphysema, while less frequent, tend to become permanent maladies. Those affected by most acute conditions recover or die without creating much of a strain on the system or leaving a trail of disability and on-going need. Chronic conditions, however, often require decades-long management, creating financial burdens for those affected and for the healthcare system.

**Box 11.9: Proportion of Deaths Attributed to Leading Causes United States 1900 and 2020**

| | *1900* | *2020* |
|---|---|---|
| Heart disease | 8.0% | 20.6% |
| Cancer | 3.7% | 17.8% |
| Tuberculosis | 11.3% | * |
| Diarrhea | 8.1% | * |
| Chronic lower respiratory diseases | * | 5.6% |
| Stroke | 4.5% | 4.7% |
| Liver disease | 5.2% | 1.5% |
| Accidents | 5.1% | 5.7% |
| Alzheimer's disease | * | 4.0% |
| Diabetes | * | 3.0% |
| Pneumonia/influenza | 10.2% | 1.6% |
| Nephritis and related | * | 1.6% |
| Suicide | * | 1.3% |

*Negligible proportion

*Note* These figures do not include COVID-19 deaths for 2020 since that was considered an anomaly. However, it should be noted that COVID-19 accounted for over 345,000 deaths in 2020, making this malady the third leading cause of death

*Source* National Center for Health Statistics

The rise of chronic conditions means that, as a population, we are probably sicker than we were a generation or two ago. In fact, the proportion of the population with at least one chronic condition has steadily increased since the 1990s, and the proportion with multiple chronic conditions is increasing apace. It seems we are trading shorter life expectancy for more years but with the additional years characterized by a lower quality of life. Box 11.10 discusses the contention that Americans are becoming sicker.

**Box 11.10: Americans Are Getting Sicker**

Americans have benefited from a 100-year run of improving health status. During the twentieth century we saw dramatic decreases in mortality (particularly maternal and infant mortality) accompanied by significant increases in life expectancy (from less than 50 years in 1900 to around 80 years today). The introduction of public health measures, advances in medicine and, importantly, increasing standards of living combined to create a population that was bigger, stronger and healthier.

There is a growing body of evidence that all of this is changing—that Americans are, in fact, getting sicker. Few observers were initially willing to state unequivocally that the health status of our population was declining. Now, however, there is ample evidence that something has gone awry when it comes to the health of Americans.

Some of the trends observed were not totally unexpected. Declines in the death rate cannot continue indefinitely nor can life expectancy continue to increase at the same rate as in the past. However, the leveling off of these trends is uniquely American, with the rates for comparable countries continuing to improve. Even worse, however, is that the rates for some subgroups within the population have actually reversed direction with some experiencing increasing mortality and declining life expectancy—trends virtually unprecedented in modern industrial societies.

Since mortality is less salient measure of population health today than in the past, social epidemiologists now focus more on morbidity rates—that is the level of sickness and disability within the population. The US is experiencing a resurgence in the incidence of many diseases once thought relegated to the "dustpan of history". While the numbers are not large to this point, it is nevertheless concerning to see an uptick in cases of measles, whooping cough, tuberculosis and malaria. The re-emergence of these conditions can be credited to a number of factors—reduced public health vigilance, lack of access to care for high-risk populations, the upsurge in immigration from countries where these conditions thrive and the anti-vaccination movement.

Although the incidence of acute conditions is an issue, it is chronic diseases that are the scourge of today's population. These "diseases of civilization" represent the predominant health threats and account for 80% of the nation's deaths. The increase in the prevalence of chronic conditions has not been a surprise to epidemiologists and represents to a great extent a natural process accompanying the aging of the US population.

An aging population, however, is not the only explanation for the surge in chronic disease. Much of the increase can be attributed to changes in the lifestyles of the US population. We have been transformed from an active population with relatively healthy eating habits to a sedentary population raised on fast food. The obesity rate for the population has exploded with a third or more of the population found to be obese. Abetted by unhealthy activities such as smoking, drinking and drug use, chronic disease now affects more than 50% of the population.

The increase in the prevalence of chronic disease first noted in the last couple of decades of the twentieth century was not totally unanticipated and, at first, was not considered cause for concern. However, the speed at which the prevalence rate increased was surprising and led investigators to examine morbidity patterns in more depth. They found that the rate of chronic disease was increasing faster than was warranted by the aging of the population. Contemporary seniors are living longer on average than seniors a generation ago, but with higher rates of chronic disease and disability. Longer life, it seems, is not being accompanied by a higher quality of life.

## *Migration*

International Immigration

As illustrated by Box 11.11, America has once again become a nation of immigrants, with the numbers of newcomers from foreign lands during the period between 1990 and 2019 equaling historical highs. The increased volume of immigrants was also noteworthy due to the regions of origin. Unlike the earlier waves of immigrants that bolstered the US population around the turn of the twentieth century, today's immigrants represent different regions and different cultures. The Europeans who had dominated the migration stream for decades were reduced to 8% of the total immigrants by twenty-first century. Their dominant role was usurped by immigrants from Latin America and Asia, accounting for 42% and 39% of the immigrants, respectively, between 2011 and 2020. Even immigrants from Africa (accounting for 10% of the total) surpassed the European numbers. In addition, long established ethnic and racial minorities are growing at faster rates than the native-born whites. Current estimates reveal an America that is becoming less "white", with percentage increases noted in the African American, Asian-American, and American Indian/Alaskan Native populations. Box 11.11 provides information on trends in immigrants by world region over the past 90 years.

**Box 11.11: Immigrants by Continent of Origin United States 1931–2020**

| *Decade* | *Total immigrants*[a] | *Europe* | *Asia* | *Other America* | *Africa* |
|---|---|---|---|---|---|
| 2011–2020 | 10,928 | 836 (8) | 4053 (39) | 4304 (42) | 1042 (10) |
| 2001–2010 | 10,501 | 1165 (8) | 4088 (29) | 4510 (43) | 926 (7) |
| 1991–2000 | 9080 | 1309 (14) | 2890 (32) | 4449 (49) | 382 (4) |
| 1981–1990 | 7256 | 706 (10) | 2814 (39) | 3581 (49) | 192 (3) |
| 1971–1980 | 4493 | 801 (18) | 1634 (36) | 1929(43) | 92 (2) |
| 1961–1970 | 3322 | 1239 (37) | 445 (13) | 1579(48) | 39 (1) |
| 1951–1960 | 2516 | 1492 (53) | 157 (6) | 841(33) | 17 (1) |
| 1941–1950 | 1035 | 622 (60) | 59 (3) | 355(34) | 7 (0) |
| 1931–1940 | 528 | 348 (66) | 15 (3) | 160(30) | 2 (0) |

[a]Numbers in thousand, percentages in parentheses
- Less than 1000
*Source* US Census Bureau

The census has documented the rapid growth of the Hispanic population and, by the beginning of the twenty-first century, Hispanics had surpassed African Americans as a percentage of the population. Since most of the population growth during the next two decades will be a function of immigration, the proportion of non-Hispanic whites within the population will continue to decline. (A telling statistic is the fact that, by 2008, racial and ethnic minorities accounted for over 50% of the children under 5, but less than 40% of the total population.) Box 11.12 illustrates the extent of the impact of immigrants on US population growth.

**Box 11.12: Immigrants and the Proportion of Population Growth Due to Immigration by Decade for the United States 1831–2020**

| *Decade* | *Immigrants (in thousands)* | *Population growth for Decade (in thousands)* | *Percent of population Growth due to Immigration* |
|---|---|---|---|
| 2011–2020 | 8350 | 20,800 | 40.1 |
| 2001–2010 | 13,863 | 27,324 | 50.9 |
| 1991–2000 | 9080 | 32,712 | 27.8 |
| 1981–1990 | 7258 | 22,164 | 32.7 |
| 1971–1980 | 4493 | 23,244 | 19.3 |
| 1961–1970 | 3322 | 23,979 | 13.9 |
| 1951–1960 | 2515 | 27,767 | 9.1 |
| 1941–1950 | 1035 | 19,028 | 5.4 |
| 1831–1840 | 599 | 4203 | 14.3 |
| 1931–1940 | 528 | 8894 | 5.9 |
| 1921–1930 | 4107 | 17,064 | 24.1 |
| 1911–1920 | 5736 | 13,738 | 41.8 |
| 1901–1910 | 8795 | 15,978 | 55.0 |
| 1891–1900 | 3688 | 13,047 | 28.3 |
| 1881–1890 | 5247 | 12,792 | 41.0 |
| 1871–1880 | 2812 | 10,337 | 27.2 |
| 1861–1870 | 2315 | 8375 | 27.6 |
| 1851–1860 | 2598 | 8251 | 31.5 |

*Source* US Census Bureau

Internal Migration

In the US today, migration is the most dynamic and complex of the three population processes, as well as the most difficult to measure. About 10% of the population changes residence each year (down from 20% in the 1940s). Some 60% are classified as movers (within county) and 38% are classified as migrants (crossed a county boundary. Of the nearly 16 million annual internal migrants in 2021, about 9 million moved within the state of previous residence and 7 million moved across a state boundary. Despite the slowing of the rate of internal migration, this process remains the most significant component of population change. Internal migration will continue to reconfigure the US population distribution.

***The most significant implication of recent internal migration has been the redistribution of population among the nation's four regions.***

The most significant implication of recent internal migration has been the redistribution of population among the nation's four regions. The general flow of population in the US since the colonization of the new world by Europeans has been from east to west. The country's population center at the time of the first census in 1790 was on the east coast; today, the population center is in Missouri.[1]

As the country became fully settled, the general trend toward east-to-west movement continued but a more complicated picture emerged. During the first half of the twentieth century a large number of people left the South, destined primarily for the Midwest and the Northeast but also contributing to the influx of new residents into the West region. This flow primarily involved African Americans who had been displaced from farm work with the mechanization of agriculture. Until the turn of the twenty-first century the east and west coasts were considered by the national elite as the only important parts of the country. Everything in between was considered "fly-over" country with little significance for the nation.

***Since 1995 the highest levels of both in- and outmigration of all four census regions occurred in the South.***

Today, the South has more population than any of the other three regions. Since 1995 the highest levels of both in- and outmigration of all four census regions occurred in the South. During this period the South gained 20 people through migration for every 1000 individuals living there in 1995. The Northeast and the Midwest displayed different migration patterns from the South, experiencing significant population loss for decades. Although the second-highest level of in-migration has been experienced by the West, in recent years that region has reported significant out-migration.

A major implication of these trends is that we are becoming a nation of Southerners and Westerners, a development representing much more than the physical reshuffling of the US population. For two or more centuries, the tenor of American society has been set by residents of the east coast, particularly the Northeast region (including the New England states, New York and New Jersey). Over time, the Midwest region began to assert its influence, with many thinking of the Midwestern states as the epitome of "middle America" (in contrast to effete east coast denizens). Twenty-five years ago social commentators predicted that we would all talk like Midwestern newscasters by the twenty-first century.

[1] The "population center" refers to the point at which, if the United States were a plate with each person standing on, it would balance.

Today, the influence of the South and West cannot be overlooked. The South is now the largest region in terms of population, and its unique characteristics make it stand out from the other regions in many ways. Although increasingly peopled by residents who were born outside the South, it continues to exert an influence that reflects its regional character. The differences between the regions that were predicted to disappear by the twenty-first century have, in fact, become more pronounced if anything.

***A major trend since World War II has been the shift in residence based on type of community.***

The other major trend since World War II has been the shift in residence based on type of community. At the time of the first census in 1790, 95% of the population lived in what today are classified as "rural" areas. With the advent of industrialization in the US a mass movement from rural areas to urban areas occurred, to the point that today only 16% of US citizens live in communities that are considered rural. The rural-to-urban migration trend peaked in the 1970s and a new flow emerged—from urban areas to suburban areas. Since that period the major flow has been out of the nation's central cities and into surrounding suburbs.

One of the consequences of these trends in internal migration is the creation of a nation of suburbanites. The predominance of urban living over rural living was well established by the middle of the twentieth century. Most of the post-WWII home building of necessity took place in suburban fringes of existing cities. At the same time, the negative aspects of urban life drove people to the suburbs during the 1960s, 70s and 80s. By the 1990s more people lived in suburbs than in any other community type. Rural areas continued to lose residents to both urban and suburban communities, but the major losers have been the central cities that saw their centers being carved out as the surrounding suburbs prospered. The central city came to be seen in many cases as the "hole" in the donut, a comparatively empty center surrounded by teeming suburbs.

***The migration patterns of the late twentieth century turned America into a nation of suburbanites.***

By the turn of the twenty-first century the suburban movement had slowed. Most people who were candidates for movement to the suburbs had done so, and, in fact, many suburbs were coming to resemble the old urban centers in terms of their demographic and social characteristics. At the same time, growing numbers of Americans had become disenchanted with the suburbs, feeling that they were not delivering on the promised benefits. Fewer households with children were being established, and these smaller households had less need or desire for their own patch of land.

Transportation costs and other economic considerations also encouraged some residents to reconsider urban living and forego the investment (and financial risk) associated with suburban living. The Millennial generation appears to be adding fuel to this fire as fewer of them are driving and increasing numbers do not have the interest or financial resources to invest in a suburban home. By 2012 the growth rate for urban areas exceeded that for suburban areas, suggesting a reversal of a decades-long trend.

### *Compositional Variables*

Age. The last two decades have seen the US grow steadily older. Since 2000, the national median age—the point at which one-half the population is older and one-half younger—has increased by 3.4 years, bringing it to 38.8 years in 2021. The median age for most states also increased from 2020 to 2021, indicating that population aging was essentially across the board. Continued low birth rates and the aging of the baby boom and Generation X assure an increasing median age.

Only one state's population—Maine—became slightly younger, as its median age decreased from 44.8 in 2020 to 44.7 in 2021. Although its median age decreased, Maine remained the state with the oldest median age in the nation.

**Box 11.13: Median Age by County United States 2021**

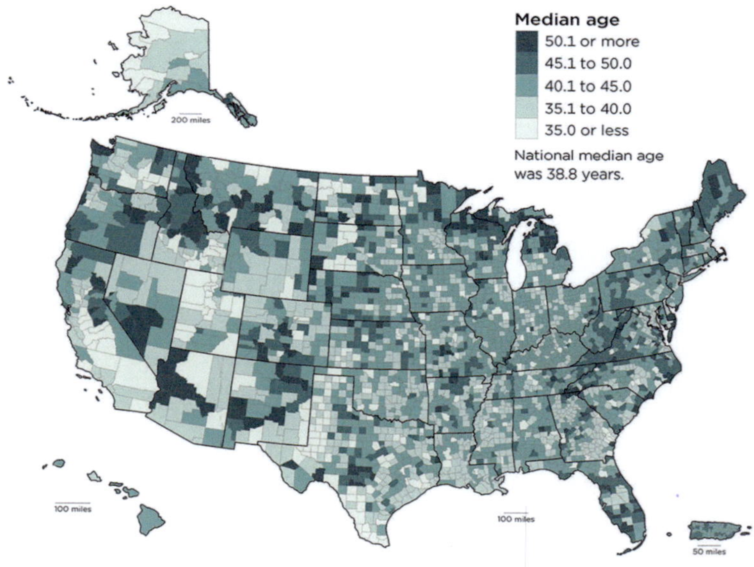


*Source* US Census Bureau

During that same period, 47 states experienced an increase in median age. Montana (40.1), New Hampshire (43.0) and West Virginia (42.8) were the only states that had no change in median age. Regionally, the Northeast was the oldest in 2021 with a median age of 40.4, followed by the Midwest (39.0), the South (38.6) and the West—which experienced the largest increase, 0.3 years to 37.7 from 2020 to 2021.

**Box 11.14: Trends in Population Proportion by Age Group United States 1900–2060 Projections**

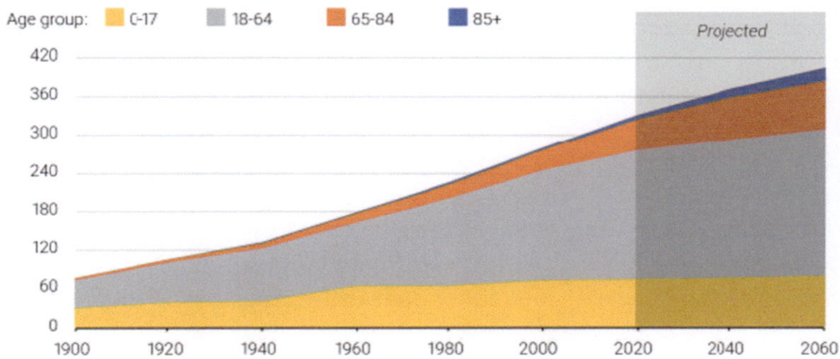


*Source* US Census Bureau

Sex Ratio. Since WWII women have been increasing as a portion of the US population. Historically, there were more men in the population than women due to the impact of maternal mortality. Younger populations have proportionally more males than females and more women in their childbearing years. In 1950, the male-to-female ratio in United States was at its highest point (99.6) and had decreased to a historic low by 1989. The sex ratio in 2020 was 97.9 males per 100 females. The percentage of population that was female was 50.5%s compare to 49.5% for males. This meant that the US had 3.43 million more females than males at that time.

As of 2020, the US population had more males than females aged below 39 years and in the 50–54 age group. The male-to-female ratio in US shifts dramatically for older populations above 60 years. At age group 60–64 years, there are five less men per 100 women. Women outnumber men by a ratio of 2-to-1 in age group 90–94 years and ratio of 4-to-1 for centenarians.

Race and Ethnicity. The cumulative effect of the trends of the past several years have resulted in a diminishing of the relative size of the non-Hispanic white population) and the growing significance of the African American, Asian and Hispanic components. Current estimates reveal an America that is becoming less "white", with percentage increases noted in the African-American, Asian-American, and American Indian/Alaskan Native populations. More importantly, the census documented the rapid growth of the Hispanic population and, by 2001, Hispanics had surpassed African Americans as a percentage of the population. Since most of the population growth during the next two decades will be a function of immigration, the proportion of non-Hispanic whites within the population will continue to decline.

The result of these trends is that the US population is becoming increasingly racially and ethnically diverse. In 2020 70.4% of the population was racially classified as white. Another 12.6% were classified as black or African American, 5.6% as Asian (along with 0.2% Hawaiian/Pacific Islander), and 0.8% as American Indian/Alaskan native. The proportion classified as white is somewhat misleading, since 19% of these are also classified as Hispanic. That means in 2021 non-Hispanic whites constituted 62.2% of the population (in contrast to the 73.7% white). Between 2010 and 2020 all racial and ethnic minority groups gained population, while the non-Hispanic white population declined by 5.1 million people. (Refer to Box 4.3 for additional information.)

Household Structure. Another demographic development characterizing US society is its changing household and family structure. For decades, the family has been undergoing change. First it was high divorce rates, then it was less people marrying (and those who did marrying at a later age); then it was less people having children (and those that did having fewer of them and at a later age).

In 2020, the Census Bureau reported that 49% of the US population age 18 and over was married, a low figure compared to the 72% of adults that were married in 1960. In 2020 33% had never married, 11 were divorced, and 6% were widowed. These figures for the non-married all represent record highs and suggest a social milieu much different than that of the twentieth century.

These changes in marital status have had major implications for the US household structure. It has meant that what is popularly considered the "typical" American family (with two parents and x number of children) has become a rarity. According to the American Community Survey, only 18% of the households in 2020 fell into this category. Today, married couple (without children) households have become the most common household form, but this type of household accounts for less than 29% of the total. "Non-traditional" households have become the norm, and an unprecedented proportion of households are one-person households. Box 11.15 presents changes is US household structure from 1990 to 2020.

**Box 11.15: Household Composition of the United States: 1990 to 2020 (in Thousands)**

| | *1990* | *2000* | *2010* | *2020* |
|---|---|---|---|---|
| Number of Households | 93,347 | 104,705 | 114,567 | 123,600 |
| Persons per household | 2.63 | 2.62 | 2.63 | 2.53 |
| Persons per family household | 3.16 | 3.14 | 3.23 | 3.15 |
| Family households | 70.8% | 68.8% | 66.4% | 67.7% |
| Married couple | 56.0% | 52.8% | 48.6% | 48.1% |

| | | | | |
|---|---|---|---|---|
| Married couple with children | 25.2% | 23.5% | 20.2% | 19.0% |
| Female-headed households (no spouse) | 13.5% | 12.1% | 13.1% | 12.0% |
| Nonfamily households | 29.2% | 31.2% | 33.6% | 32.3% |
| One-person households | 12.4% | 14.0% | 27.4% | 29.2% |

*Source* US Census Bureau

The changes in marital status and household structure have been so significant that the Census Bureau has had to develop a new typology of household types in order to properly classify the population. This typology has to consider in addition to "traditional" household types (e.g., nuclear families, extended families) new categories of blended families, "stem" families (with only one parent and a related child), cohabitating families and so forth. Thus, a child residing in a household could be the biological child of one parent or two parents (married or not), a stepchild of one or both parents, a grandchild of one or both parents, an adopted child, an unrelated child living in the household, and so forth.

The large number of births taking place outside of marriage, of course, has served to dramatically alter family and household structures. The fastest growing household type over the past two decades has been single-parent-with-children households. In addition, many of the women having children outside of marriage remain in *their* parents' households, creating extended families out of necessity rather than choice. This has also led to a high and growing proportion of children being raised by grandparents or some other relative. While some of the arrangements have been present within African American communities for some time, they are now becoming manifest in white and Hispanic communities.

***The large number of births taking place outside of marriage has served to alter family and household structures.***

Another trend that represents even more of a break with the past is the number of cohabitating households. Recent research finds that 65% of women 19–44 lived in cohabitating relationships prior to marriage. This compares to less than 10% in 1970. Further, some 25% of those in cohabitating households bore children while in the relationship. The cohabitation rates were higher for African Americans and Hispanics than for non-Hispanic whites.

All of the developments taken together have resulted in an America that is quite different from that experienced by previous generations. Box 11.16 illustrates the extent of the change by describing the various "firsts" that characterize contemporary American society.

**Box 11.16: Demographic "Firsts" in America**
There are a number of trends that are occurring within the US that are noteworthy. These include:

- A rate of fertility so low that the US population cannot reproduce itself
- The dramatic aging of the population
- The increasing racial/ethnic diversity of the population
- The changing family structure and increase in non-traditional households
- An unprecedented increase in death rates and the decline in life expectancy
- The deteriorating health of the US population

The changes that are occurring of such significance that we as a population are constantly setting new records. Today, for example, the American population is characterized by:

- The largest elderly population we have ever had (both numbers and percent)
- The greatest "excess" of women we have ever had
- The lowest marriage rate we have ever had
- The largest group of adults we have ever had who have never been married (both numbers and percent)
- The most births outside marriage we have ever had
- The most single-parent households we have ever had (both numbers and percent)
- The most women working outside the home that we have ever had
- The highest level of immigration (both legal and illegal) we have ever had

As a result of these developments our American "portrait" in the early twenty-first century looks much different than it did only a couple of decades ago.

## Key Points

- The demographic developments occurring over the past 25 years are not extensions of previous trends but are unique in our history and unprecedented in their impact.
- Compared to our own historical rates and to those of much of the rest of the world, the level of US fertility is quite low today.
- While Americans have been recording relatively high numbers of births, they are also recording record low fertility rates.
- In 2020, nearly 41% of all births in the United States were to unmarried mothers, representing a dramatic increase over previous generations.
- The long-term fertility pattern for the US reflects a major reversal of historical patterns.
- Women entered the workforce in unprecedented numbers beginning in the 1960s, thereby making a major contribution to declining fertility rates.

- The maternal mortality rate for the US has reversed course and is increasing to a level not seen in decades.
- Significant changes have occurred with regard to the major causes of death affecting US society.
- The rise of chronic conditions means that, as a population, we are probably sicker than we were a generation or two ago.
- The most significant implication of recent internal migration has been the redistribution of population among the nation's four regions.
- The South is now the largest region in terms of population, and its unique characteristics make it stand out.
- A major trend since World War II has been the shift in residence based on type of community.

- The most significant implication of recent internal migration has been the redistribution of population among the nation's four regions.
- The South is now the largest region in terms of population, and its unique characteristics make it stand out.
- A major trend since World War II has been the shift in residence based on type of community.

- The migration patterns of the late twentieth century turned America into a nation of suburbanites.
- The large number of births taking place outside of marriage has served to alter family and household structures.

## Discussion Questions

11.1 The demographic attributes of American society have changed significantly over the past quarter of century. As you consider these changes, do you feel that society has changed for the better or for the worse? Explain your thinking.

11.2 Like many developed countries, the US is facing a "demographic crisis". It is not reproducing fast enough to maintain population growth. Discuss the factors that are contributing to this "crisis" and brainstorm ways to encourage population growth.

## Exercises

*Exercise 11.1: The Consequences of Current Demographic Trends*

A number of demographic trends that are currently affecting the US population are presented below. Review each of the trends and discuss the implications of these trends for our **economy**. Consider the ways in which these trends are affecting or will affect the types of workers that we can expect to be available, the characteristics of consumers, the types of goods and services that will be required in the future, and any other consequence related to the economy that might result from these trends. Be prepared to explain your conclusions to the rest of the class.

- The US population is rapidly aging as the baby boom cohort moves into its senior years, the elder population continues to grow, and the working age and young adult populations continue to decline.
- The US population is becoming more "feminized" as the population ages and women constitute an ever-growing proportion of the population.
- The US population is becoming increasingly diverse in terms of race and ethnicity, with the Hispanic population growing dramatically and significant increases in a number of other ethnic groups.
- The "traditional" family is becoming less common in US society as fewer people marry and childless couples, single-parent households, cohabiting (sometimes gay) couples, and individuals living alone become more common.
- The US population is becoming more diverse in terms of religion but less religious overall in terms of formal religious participation.
- The unemployment rate is at a modern-day high level and is likely to continue at an elevated level for the foreseeable future as more jobs are shipped overseas and domestic industry is slow to recover from the recession.
- The income gap between the "haves" and the "have nots" in the US is increasing and the poverty population has begun to grow after experiencing more than two decades of decline.

## References

Daniels, K., & Abma, J. C. (2020). Current contraceptive status among women aged 15–49: United States, 2017–2019. *NCHS Data Brief, 388*

Ezzati, M., Friedman, A. B., Kulkarni, S. C., & Murray, C. J. L. (2008). The reversal of fortunes: Trends in county mortality and cross-county mortality disparities in the United States. *PLoS Medicine*. Downloaded from https://journals.plos.org/plosmedicine/article?id=10.1371/journal.pmed.0050066.

### Additional Resources

Goodhart, C., & Pradhan, M. (2020). *The great demographic reversal*. New York: Springer.

Hamish, L. (2015). *Demographic trends in the United States. The changing electoral demographics*. GRIN Vertag. Munich, Germany.

Singelmann, J., & Poston, D. (2020). *Developments in demography in the 21st century*. New York: Springer.

CHAPTER 12

# World Population Trends

**Abstract** This chapter extends the examination of demographic trends in the US to the world at large. The trends affecting the world's population are explored from a global perspective and the extent to which they will shape the world of the future considered. The worldwide impact of fertility, mortality and especially migration trends are discussed, and speculation offered on the future size and characteristics of the world's population.

## Introduction

The world's population is constantly undergoing change and the demographic developments expected in coming years are likely to be significant—especially as the world faces a number of social, economic, political and environmental challenges. This chapter reviews population trends at the global, regional and national levels and reports on key indicators commonly used by the United Nations system for monitoring demographic change.

## Population Growth and Change[1]

Much has been made of the size of the world's population and the factors that are influencing population growth. The UN has developed three population projections, called variants—one high, one low, and one in between. The medium projection suggests that the world population will grow to nearly 10 billion by the middle of the twenty-first century and level off at around 11

[1] The manuscript was prepared during the global COVID-19 pandemic and figures presented here should be viewed in light of the potential impact of the pandemic on demographic trends.



R. K. Thomas, *Demography: An Introduction to Population Studies*, Springer Texts in Social Sciences, https://doi.org/10.1007/978-3-031-56623-3_12

billion by the end of it. However, if fertility declines by less than assumed, the world population could grow to as many as 16.5 billion by the end of the century. If fertility falls by more than assumed, the world population could actually fall below the current level of about 8 billion.

According to UN figures, the world population reached 7.8 billion in 2020. This means the world added approximately one billion people between 2010 and 2020. Sixty percent of the global population lives in Asia, 17% in Africa, 10% in Europe, 8% in Latin America and the Caribbean, and the remaining 5% in Northern America and Oceania. China and India remain the two most populous countries with more than 1 billion people each, representing 19% and 18% of the world's population, respectively. The US population comes in a distant third with 335 million residents (or 4% of the world's population).

Among the ten largest countries in the world, one is in Africa (Nigeria), five are in Asia (Bangladesh, China, India, Indonesia, and Pakistan), two are in Latin America (Brazil and Mexico), one is in Northern America (the United States), and one is in Europe (Russian Federation). Amongst these, Nigeria's population, currently the seventh largest in the world, is growing the most rapidly. Consequently, the population of Nigeria is projected to surpass that of the United States by about 2050, at which point it would become the third largest country in the world. Box 12.1 presents current and projected populations by world region according to the United Nations.

**Box 12.1: Population for the World and Major Subdivision 2020 and 2050 Projected (in Millions)**

| | *2020* | *Projected* |
|---|---|---|
| World | 7801 | 9825 |
| Africa | 1341 | 2489 |
| Asia | 4641 | 5290 |
| Europe | 748 | 710 |
| Latin America and the Caribbean | 654 | 762 |
| Northern America | 369 | 425 |
| Oceania | 43 | 57 |

*Source* United Nations

General population trends mask considerable differences among individual countries. On the one hand, there are the world's least developed countries, which continue to have high fertility levels. According to conservative projections, the combined population of the world's least developed countries (LDCs) will double by 2050, and in some countries it will even triple. On the other hand, there are high-income and rising-income countries, which are experiencing slow population growth or no population growth at all. Whereas

LDCs will continue to have large, growing populations of young people, the latter have large, growing populations of older people. There are even some countries experiencing an overall decline in population. The sections below describe the role of various contributors to population change. Box 12.2 illustrates the distribution of the world's population.

**Box 12.2: Distribution of the World's Population 2020**

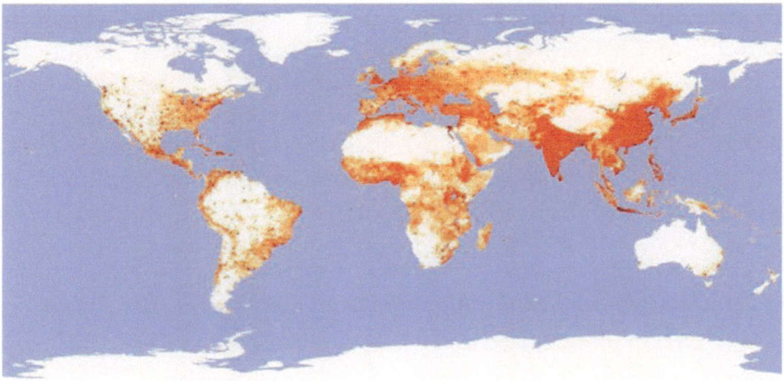

*Source* United Nations

### *Components of Population Growth and Change*

The huge growth in the world population over the past two centuries is largely the result of advances in modern medicine and improvement in living standards. These have significantly reduced infant, child and maternal mortality, contributing to an increase in life expectancy. Although fertility levels have declined, they have not fallen at the same pace as mortality levels. (Remember the "demographic transition" discussed earlier.)

***It is estimated that 83 million people are added to the world's population annually.***

Fertility. Recent years have witnessed enormous changes in fertility rates worldwide. In the early 1970s, women had on average 4.5 children each; by 2020, total fertility for the world had fallen to around 2.4 children per woman. Future population growth is highly dependent on the path that future fertility will take, as relatively small changes in fertility behavior, when projected over several decades, can generate large differences in total population. In the medium-variant projection, global fertility declines from 2.5 children per women in 2010–2015 to 2.25 children per women in 2045–2050 and 2.0 children per women in 2095–2100. If fertility were to be consistently half a child above those levels, world population would reach 10.8 billion by 2050 and 16.6 billion by 2100. Fertility levels consistently half a child below the medium variant would lead to a population of 8.7 billion by mid-century and 7.3 billion by 2100.

In Africa, where fertility levels are the highest of any major area, total fertility has fallen from 4.9 children per woman in 2005–2010 to 4.7 children per woman in 2010–2015. Fertility levels have also fallen in Asia and Oceania over the same period, from 2.3 to 2.2 children per woman in Asia and from 2.5 to 2.4 children per woman in Oceania. Recent fertility declines have been slightly larger in Latin America and the Caribbean where fertility has fallen from 2.3 to 2.2 and in Northern America where fertility has fallen from 2.0 in 2005–2010 to 1.89 in 2010–2015. Europe is the only major area that was an exception to this trend. In recent years, total fertility in Europe has increased slightly from 1.55 children per woman in 2005–2010 to 1.6 children per woman in 2010–2015.

***Future population growth is highly dependent on the path that future fertility takes.***

The 48 least developed countries (LDCs) as a whole still have high total fertility (4.3 children per woman in 2010–2015) and fast-growing populations at 2.4% per year. Although this rate of increase is expected to slow significantly over the next decades, the population of the LDCs (954 million in 2015) is projected to increase 39% between 2015 and 2030, and to double to 1.9 billion persons by mid-century.

Mortality. Mortality rates worldwide continue to drop, with a commensurate increase in global life expectancy. Under-five mortality, expressed as the probability of dying between birth and a child's fifth birthday, is an important indicator of development and the well-being of children. Globally, deaths among children under age five fell from 71 per 1000 live births in 2000–2005 to an estimated 50 per 1000 in 2010–2015. Absolute declines in rates were particularly large in Sub-Saharan Africa (142–99 per 1000) and in the least developed countries (125–86 per 1000).

The reduction of under-five mortality has proceeded swiftly in many countries in the past decade. In the majority of countries in Sub-Saharan Africa and the group of LDCs, the annual pace of decline in under-five mortality accelerated after 2000.

Significant gains in life expectancy have been achieved globally in recent years. Life expectancy at birth rose by three years between 2000–2005 and 2010–2015 (that is from 67 to 70 years. All major areas shared in the life expectancy gains over this period, but the greatest increases were in Africa, where life expectancy rose by 6 years in the 2000s after rising by only 2 years in the previous decade. Life expectancy in Africa in 2010–2015 stood at 60 years, compared to 72 years in Asia, 75 years in Latin America and the Caribbean,

77 years in Europe and Oceania and 79 years in Northern America. Africa is projected to gain about 19 years of life expectancy by the end of the century, reaching 70 years in 2045–2050 and 78 years in 2095–2100. Such increases are contingent on combating infectious as well as non-communicable diseases. Both Asia and Latin America and the Caribbean are projected to gain 13–14 years of life expectancy by 2095–2100, while Europe, Northern America and Oceania are projected to gain 10–11 years. Box 12.5 presents a population pyramid for the current world's population.

Migration. Internal and international migration can be positive forces for economic and social development. These processes provide a mechanism for rebalancing labor markets in areas of origin and destination and accelerating the diffusion of new ideas and technologies. Migration can also result in significant flows of remittances to areas of origin. Historically, international migration has been a much smaller component of population change than births or deaths. However, in some countries and areas the impact of migration on population size is significant, especially for countries that send or receive proportionately large numbers of economic migrants or those affected by refugee flows.

Overall, between 1950 and 2015, the major areas of Europe, North America and Oceania have been net receivers of international migrants, while Africa, Asia and Latin American and the Caribbean have been net senders. From 2000 to 2015, average annual net migration to Europe, Northern America and Oceania averaged 2.8 million persons per year.

When countries are grouped by income rather than geography, the attraction of high-income countries is even more evident: from 2000 to 2015, high-income countries received an average of 4.1 million net migrants annually. In the future, net migration is projected to be a major contributor to population growth in many high-income countries. Box 12.3 depicts current world migration patterns.

**Box 12.3: World Migration Patterns**

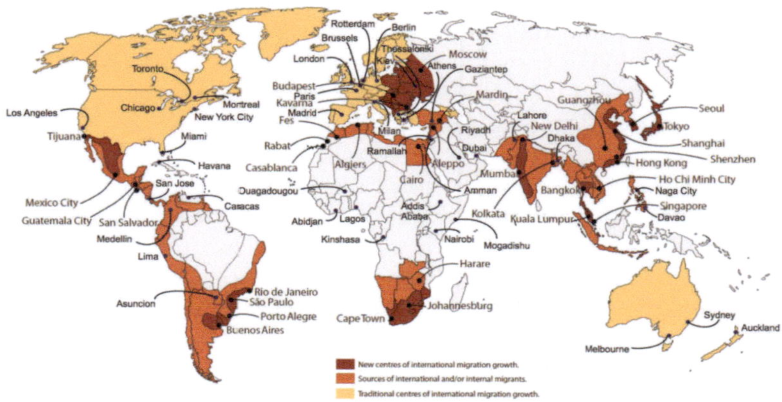


*Source* United Nations

In 2022, the number of refugees around the world exceeded 30 million—an escalation of what was already the highest number of refugees on record. In the past decade, the global refugee crisis has more than doubled in scope. In 2022, the United Nations High Commission on Refugees (UNHCR) announced that the world had surpassed the 100 million mark for total displacement, meaning that over 1.2% of the global population have been forced to leave their homes. Among these people are over 32.5 million refugees. Some 76% of those refugees come from just six countries.

Due to improved survival rates and past high fertility levels world population will continue to grow for decades to come. "Population momentum" will exist because there are more women of reproductive age today than in the past. This will contribute to a relatively large number of births, even if those women have fewer children on average. Although population growth is currently primarily attributable to population momentum, after 2060 it will almost exclusively be driven by fertility levels in the world's least developed countries.

While continued population growth presents challenges, the move from higher to lower mortality and fertility reflects achievements in health, education and human rights. Falling fertility levels also create opportunities for countries to realize a "demographic dividend"—the economic growth that can occur when there is an increase in the number of people of working age and a decrease in the number of dependents.

Although the growth rate of the "least developed countries (LDCs) is projected to slow from its current 2.4% annually, the population of this group is projected to double in size from 954 million inhabitants in 2015 to 1.9 billion in 2050 and further increase to 3.2 billion in 2100. Between 2015 and 2100, the populations of 33 countries, most of them LDCs, have a high probability of at least tripling. Among them, the populations of Angola, Burundi, Democratic Republic of Congo, Malawi, Mali, Niger, Somalia, Uganda, United Republic of Tanzania and Zambia are projected to increase at least five-fold by 2100.

***Continued population growth until 2050 is almost inevitable, even if the decline in fertility accelerates.***

In sharp contrast, the populations of 48 countries or areas in the world are expected to decrease between 2015 and 2050. Several countries are expected to see their populations decline by more than 15% by 2050, including Bosnia and Herzegovina, Bulgaria, Croatia, Hungary, Japan, Latvia, Lithuania, Republic of Moldova, Romania, Serbia, and Ukraine. Fertility in all European countries is now below the level required for full replacement of the population in the long run (around 2.1 children per woman), and in the majority of cases, fertility has been below the replacement level for several decades. See Box 12.6 for a discussion of the demographic makeup of the world's population.

More than half of global population growth between now and 2050 is expected to occur in Africa. Africa has the highest rate of population growth among major areas, growing at a pace of 2.6% annually in 2010–2015. Consequently, of the additional 2.4 billion people projected to be added to the global population between 2015 and 2050, 1.3 billion will be added in Africa. Asia is projected to be the second largest contributor to future global population growth, adding 0.9 billion people between 2015 and 2050, followed by Northern America, Latin America and the Caribbean and Oceania, which are projected to have much smaller increments. In the medium variant, Europe is projected to have a smaller population in 2050 than in 2015.

These megatrends have far-reaching implications. They affect economic development, employment, income distribution, poverty and social service needs. They also affect efforts to ensure universal access to healthcare, education, housing, sanitation, water, food and energy. To more sustainably address the needs of individuals, policymakers must understand how many people are living on the planet, where they are, how old they are, and how many people will come after them.

While demographics vary considerably at national levels, the overall trends have global implications for sustainable development. The global climate will change no matter where greenhouse gases are emitted, for example. Efforts to sustainably meet the needs and desires of a growing world population will have implications for all countries—as will failure to meet these needs. Box 12.4 examines the special case of Russia.

**Box 12.4: The Special Case of Russia**

During the last several years, the Russian public and Russian policymakers have become increasingly concerned about demographic trends in their country. Russia is currently experiencing unusually high mortality rates from preventable causes (e.g., alcoholism), extremely high induced abortion rates, and fertility rates that are among the lowest in the world.

Add to this the toll taken by the COVID-19 pandemic (perhaps the highest in the world) and the effects of the war in Ukraine and a demographic crisis is unavoidable. It is not only the casualties of war but the 500,000 military-eligible Russian males who have fled the country.

As a result of these developments, the population growth rate in Russia has become negative (as seen in the graph below), life expectancy appears to be decreasing, and the population has been rapidly aging. The evidence now available on these changes has led a number of observers to call the present situation a "demographic crisis."

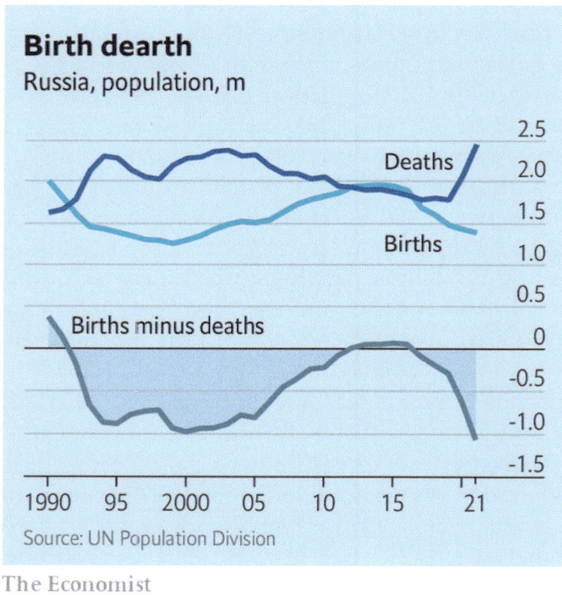


*Source* United Nations

The public and the government are concerned about the detrimental effects that demographic trends may have on economic reform and development—for example, how a shrinking working-age population will support a growing number of elderly and the demands that the latter will make on the social safety net. Policymakers and demographers in Russia are eager to learn more about the causes and likely social and economic consequences of these demographic changes so that they can define appropriate domestic policy options as the reform process proceeds.

It has been observed that the changes that the Russian family has undergone in recent decades have undermined traditional family patterns and produced alternative forms, such as single-parent families and unregistered cohabitation. The recent decrease in Russian fertility is thought to have been triggered by economic hardships.

In contrast to other countries, age at marriage in Russia has been constantly decreasing and most reproductive activity is now concentrated in the relatively early ages. Pro-natalist policy measures enacted by the Soviet government in the 1980s led to an increase in fertility rates in the mid-1980s, owing largely to a rise in the number of second births. However, this fertility increase was short-lived. As the financial advantages of the government measures were swept away by inflation and the economic hardships of the early 1990s, fertility, which in any case is rarely affected by such measures on a long-term basis, dropped to one of the lowest levels in the world. Drs. Zakharov and Ivanova concede that the present state of fertility in Russia is not necessarily anomalous and may be the normal outcome of a longtime trend that has brought Russia to the beginning of what is often called "second demographic transition"—a fertility decline to below-replacement levels associated with greater individualism and less interest in childbearing.

The dramatic worsening of the epidemiological situation and the sharp drop of life expectancy are the most unequivocal signs of the Russian demographic crisis. Today, as before, Russia holds two somber records: the lowest life expectancy among males for a developed country in peacetime and the largest gap between male and female life expectancy in the world. Recent mortality increase has been especially dramatic among working-age men. When the mortality rate is disaggregated by the cause of death, it becomes obvious that deaths from accidents, violence, poisoning, and other non-natural causes are primarily responsible for the latest rise in mortality.

*Sources* DaVanzo and Farnsworth (1996); The Economist (2023). "Russia's Population Nightmare is Going to Get Even Worse." Downloaded from URL: https://www.economist.com/europe/2023/03/04/russias-population-nightmare-is-going-to-get-even-worse

### *Changing Demographic Attributes*

Age Distribution. The median age of the global population, that is, the age at which half the population is older and half is younger, was 30.9 years in 2020. This reflects a considerable increase over previous decades. About one-third (or 33%) of the world's people was under 20 years of age in 2020, 53% were aged 20–59 years, and 14% were 60 years or over. These figures represent a slight shift upwards for the working age and elderly populations.

Populations in many parts of the world are still young. In Africa, children under age 15 accounted for 41% of the population in 2015 and young persons aged 15 to 24 accounted for a further 19%. Latin America, the Caribbean and Asia, which have seen greater declines in fertility, have smaller percents of children and similar percentages of youth. In total, these three regions were home to 1.7 billion children and 1.1 billion young persons in 2015. The proportions of children in the populations of many countries of these regions are projected to decline further in the near-term future, while the size and the proportion of populations in the prime working ages can be expected to grow.

Countries with a relatively high ratio of working to dependent populations have the possibility of benefitting from the assumed demographic dividend, provided that appropriate labor market and other policies allow for a productive absorption of the growing working-age population and for increased investments in the human capital of children and youth.

Globally, the population cohort 60 or over is the fastest growing. As fertility declines and life expectancy rises, the proportion of the population above a certain age rises. This population aging is occurring throughout the world. In 2015, there were 901 million people aged 60 or over, comprising 12% of the global population. The population aged 60 or above is growing at a rate of 3.26% per year. Currently, Europe has the greatest percentage of its population aged 60 or over (26%), but rapid aging will occur in other parts of the world as well, so that, by 2050, all major areas of the world except Africa will have nearly a quarter or more of their populations aged 60 or over. The number of older persons in the world is projected to be 1.4 billion by 2030 and 2.1 billion by 2050 and could rise to 3.2 billion in 2100. In the short-to-medium term, higher numbers of people in older age cohorts are inevitable, given that the relevant cohorts are already alive.

As the population growth rate falls over time, the proportion of older persons increases while that of younger persons decreases. In 2015, there were more than twice as many children under the age of 15 in the world as there were older persons aged 60 or above. By 2050, however, there will be almost complete global parity between the number of older persons aged 60 and above and the number of children under the age of 15.

Population aging is projected to have a profound effect on the number of workers per retiree in various countries, as measured by the Potential Support Ratio (PSR), defined as the number of people aged 20–64 divided by the number of people aged 65 and over (previously referred to as the age-dependency ratio). Currently, African countries, on average, have 12.9 people aged 20–64 for every person aged 65 or above, while Asian countries have PSRs of 8.0, Latin America and the Caribbean 7.6, Oceania 4.8 and Europe and Northern America at or under 4. Japan, at 2.1, has the lowest PSR in the world, although seven European countries also have PSRs below 3. By 2050, seven Asian countries, 24 European countries, and four countries of Latin America and the Caribbean are expected to have PSRs below 2, underscoring the fiscal and political pressures that the health care systems as well as the old-age and social protection systems of many countries are likely to face in the not-too-distant future.

In Europe, 24% of the population is already aged 60 years or over and that proportion is projected to reach 34% in 2050 and 35% in 2100. Other major areas of the world are also projected to experience significant population aging over the next several decades. For Latin America and the Caribbean, the population will be transformed from having just 11% of the total aged 60 or over in 2015 to having 26% aged 60 or over by 2050. Similarly, Asia is expected to shift from 12% aged 60 or over to 25% by 2050, Northern America from 21 to 28% by 2050, and Oceania from 16 to 23% by 2050. Africa has the youngest age distribution of any major area. Nevertheless, it is also projected to age rapidly over the next 35 years, with the percentage of its population aged 60 or over rising from 5% in 2015 to 9% by 2050.

***The number of persons aged 60 and above worldwide is expected to more than double by 2050.***

**Box 12.5: World Population Pyramid**

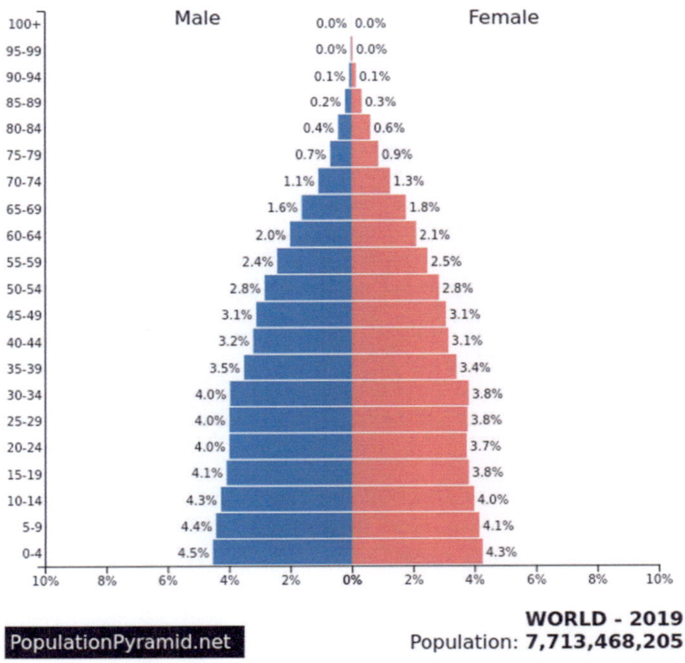


*Source* United Nations

Globally, the number of persons aged 60 and above is expected to more than double by 2050 and more than triple by 2100, increasing from 901 million in 2015 to 2.1 billion in 2050 and 3.2 billion in 2100. Sixty-six percent of the increase between 2015 and 2050 will occur in Asia, 13% in Africa, 11% in Latin America and the Caribbean, and the remaining 10% in other areas (Box 12.5).

An informative means of gaining an understanding of the nature of the world's population is to examine the distribution of the population as if the world consisted of 100 people. Box 12.6 illustrates the distribution of the world's population characteristics if viewed in this context.

**Box 12.6: If the World Were 100 People**

If we could reduce the world's population to a village of precisely 100 people, with all existing human ratios remaining the same, the demographics would look something like this:

- 60 Asians, 14 Africans, 12 Europeans, 8 Latin Americans, 5 from the USA and Canada, and 1 from the South Pacific
- 51 males; 49 females
- 82 non-whites; 18 whites
- 67 non-Christians; 33 Christians
- 80 in substandard housing
- 67 would be unable to read
- 50 would be malnourished (and 1 dying of starvation)
- 33 would be without access to a safe water supply
- 39 would lack access to improved sanitation
- 24 would not have any electricity
- 7 people would have access to the Internet
- 1 would have a college education
- 5 (all from the US) would control 32% of the entire world's wealth
- 33 would be attempting to live on only 3% of the income of "the village".

*Source* This version of the "village" was developed by Our Lady's Bishop Eton School in England (http://ourladysbishopeton.com/wp-content/uploads/2020/04/If-the-World-Were-a-Village-of-100-People.pdf). There are a number of versions, all based primarily on the following book: Smith (2020)

Family Structure. One of the consequences of these global demographic developments is a major reduction in average family size. Today, countries of the world can be roughly classified into three groups depending on their current level of fertility. It is estimated that 46% of the world's population lives in countries with low levels of fertility where women have fewer than 2.1 children, on average, over their lifetimes. Low-fertility countries now include all of Europe and Northern America, plus 20 countries of Asia, 17 in Latin America and the Caribbean, 3 in Oceania and 1 in Africa. Another 46% of the world's population lives in "intermediate fertility" countries that have already experienced substantial fertility decline. Intermediate-fertility countries are found in many regions, with the largest being India, Indonesia, Pakistan, Bangladesh, Mexico, and the Philippines. The remaining 9% of the world's population lives in high-fertility countries that have experienced only limited fertility decline to date. In these countries the average woman has 5 or more children over her lifetime. Of the 21 high-fertility countries, 19 are in Africa and 2 are in Asia.

Between 2015 and 2050, total births in the group of high-income countries are projected to exceed deaths by 20 million, while the net gain in migrants is projected to be 91 million. Thus, net migration is projected to account for 82% of population growth in the high-income countries. The movement of people from Asia, Africa and Latin America to Europe, Northern America and Oceania have dominated the world migration patterns for almost half a century, but flows among developing countries have also been important. Several high-income and middle-income countries in the "global south" have also been attracting migrants in large numbers for several years. All of these developments have implications for family structure.

## Key Points

- In 2020, world population reached nearly 8 billion.
- It is estimated that 83 million people are added to the world's population annually.
- In 2015, 9.1% of the world's population was under age 5, 26.1% was under age 15, 12.3% was 60 or over and 1.7% was 80 or over.
- Continued population growth until 2050 is almost inevitable, even if the current decline in fertility accelerates.
- Future population growth is highly dependent on the path that future fertility takes.
- The 48 least developed countries (LDCs) still exhibit high total fertility.
- The slowdown in population growth brought about by a reduction in fertility is associated with population aging.
- Africa continues to experience very high rates of population growth.
- Forty-eight countries or areas are projected to experience population decline between 2015 and 2050.
- Nine countries are expected to account for more than half of the world's projected population increase over the period 2015–2050.
- More and more countries now have below-replacement fertility and in many cases have had below-replacement fertility for several decades.
- Life expectancy at birth has increased significantly in the least developed countries in recent years.
- The six-year average gain in life expectancy among the poorest countries, from 56 years in 2000–2005 to 62 years in 2010–2015.
- Progress in reducing under-five mortality has been very significant and wide-reaching in recent years.

- A number of Eastern European countries experienced reductions in life expectancy at birth in the late 1980s and 1990s.
- From 2000 to 2015, annual net migration to Europe, Northern America and Oceania averaged 2.8 million persons per year.

## Discussion Questions

12.1 Worldwide migration is at record levels. Review current patterns of migration and consider the impact that these migration streams will ultimately have on the world's countries. Factor in the implications of tens of millions of refugees worldwide.

12.2 Review patterns of population growth and change for the various world regions and consider the implications of a situation in which some countries (typically the least developed) continue to grow in population while other countries (typically the most developed) continue to decline in population.

## Exercises

*Exercise 12.1: Population Distribution*

Using the map above and the World Online Atlas consider the following questions:

(a) Where are the regions of high population density?
(b) Where are the regions of low population density?
(c) Why is that the case?

*Exercise 12.2*

Determine the population density for the following countries using the Earth Trends database:

Bahrain
United Kingdom
Greenland
Australia
India.

Describe their population densities and discuss the implications of the level of density for the respective countries.

## Reference

DaVanzo, J., & Farnsworth, G. (Eds.). (1996). Russia's demographic "crisis". RAND Corporation. https://www.rand.org/pubs/conf_proceedings/CF124.html

## Additional Resources

Martins, J. F. G., & Swanson, D. (2018). *Global populations in transition*. Springer B.V. Press.
Smith, D. J. (2020). *If the world were a village* (2nd ed.). Kid's Can Press.
United Nations. (2019). *World demographic trends 2018*. United Nations.

CHAPTER 13

# Applied Demography

**Abstract** This chapter addresses the increasingly important field of applied demography. While demography is by definition an applied science, the subdiscipline of applied demography is particularly focused on applying demographic concepts, methods and data to real world problems. Of particular interest are such endeavors as business demography, health demography, political demography and educational demography. Concrete examples are provided of these applications and students provided an opportunity to apply their demographic insights in the consideration of real world problems.

## Introduction

Demography is by definition an *applied* discipline. Most demographic activity involves applying the concepts, methods and data to the solution of real-world problems. While there is a basic science dimension to demography—what we might call science for science's sake—most demographers use their knowledge to understand real world problems. The goal is not just to understand them, but to help develop solutions for them.

In addition to proposing practical solutions to real-world problems, demographic data make an important contribution to policy setting. An understanding of demographic trends provides policy-makers with the background they need for establishing effective policies. There is essentially no sector of US society that cannot benefit at the policy level from demographic input. In such diverse areas as education, economic development, transportation, disaster preparedness, and criminal justice, demographic data can make a significant contribution to policy setting and program development.

"Applied demography" often involves the application of demographic theories, concepts, methods and data in addressing a wide range of issues in a



R. K. Thomas, *Demography: An Introduction to Population Studies*, Springer Texts in Social Sciences, https://doi.org/10.1007/978-3-031-56623-3_13

variety of fields. This often means the application of demographic methods and materials to non-demographic factors and events. Early commentators on the nature of applied demography asserted that it is focused on prediction as opposed to description and explanation, on current and future events rather than the past, on local versus national and international demographic phenomena, and on an emphasis on the *implications* of demographic events (Rives & Serow, 1984).

***"Applied demography" involves the application of demographic theories, concepts, methods and data to the solution of practical "real world" problems.***

Perhaps the best approach to explaining applied demography would be to identify what applied demographers actually *do*. A simplified depiction of the way in which demography is applied is presented in Box 13.1.

**Box 13.1: How Demography is Applied**

**Data → Information → Policy → Action.**

Every scientific endeavor begins with data—i.e., the raw numbers that describe a phenomenon of some type (in this case a population). This would include, for example, the ages of every person in the defined population, the number of births to women in that population, or the citizenship status of all foreign-born residents. These raw *data* would essentially involve a list of numbers that do not mean much in themselves. These data must be converted into *information* that provides a description of the population in question. Thus, if we can calculate the median age for the population, the annual birth rate, or the proportion of the foreign-born who are not citizens, we now have some information we can use.

In these simplified examples, having access to information allows demographers to contribute to policy making. If, for example, the median age is 40 years the types of policies to be considered would be a lot different than if the median age were 18 years. Similarly, it makes a difference to *policy* makers if the fertility rate is 80 births per 1000 women of child-bearing age as opposed to 40. Finally, the proportion of foreign-born within a population—and their citizenship status—has implications for immigration policy and the allocation of federal funds.

The term "policy" is used loosely here to refer to any deliberations on the meaning of the statistics and any implications these statistics may have for the social, political or economic realms. An aging population, for example, raises questions about the ability of a population to replace itself. A declining fertility rate also has implications for the same issue, in that the increasing median age of the population reflects, among other things, the fact that women are having fewer children. Finally, the proportion of the population that is foreign-born has implications for immigration policy, educational resources, and the job market. The significance of the foreign-born population even has implications for the population replacement situation in that recent immigrants typically exhibit higher fertility rates than the native-born.

The ultimate "application" of demography comes at the action phase, and, for these examples, actions may include developing health services for seniors (as opposed to children), introducing incentives for more births or less (depending), and introducing English-as-a-second-language programs in public schools (or not). The impact of these actions informed by demographic analyses can be tracked over time to determine their implications for the populations in question and for society as a whole.

The relevance of demography for an understanding of the world is reflected in this sample of issues making headlines in recent years:

- Increasing income inequality as the size of the middle class dwindles
- Conflict between different ethnic and religious factions around the world
- The effects of climate change on various parts of the world
- The continuing issue of illegal immigration into the United States
- The polarization of the American electorate
- The disparate impact of the COVID-19 pandemic.

These issues all have national and/or international implications—and all are directly or indirectly related to demography. In fact, there is virtually no social, economic or political issue that does not have its roots in the demographics of the population.

While the events above have captured the headlines, there are a number of other trends occurring within the US population that are currently making headlines or reflect long-term changes in the social structure:

- The decline in the US population below replacement levels
- The aging of the US population (accompanied by growing "feminization")
- The increasing racial/ethnic diversity of the US population
- The changing family structure of the US population now including same-sex marriages)
- Increasing death rates among some segments of the US population.

These types of demographic trends have significant implications for US society present and future. The aging of the US population by itself has ushered in an unprecedented period for a society that has always emphasized its youthfulness. The dramatic increase in the Hispanic population in the US has wide-ranging implications for the economy, education, healthcare and the political system.

In view of developments like those above, applied demography seeks to interpret the political and economic events whether at the local, national or international level. We gain insights when we realize that headlines like "Growing elderly population puts pressure on Medicare," "Aging baby-boomers threaten solvency of Social Security," or "Drop in birthrate could lead to population decline" reflect the operation of demographic processes. We can better understand both the obvious and not-so-obvious dimensions of the issue if we apply demographic knowledge and techniques.

***Applied demography provides a means of interpreting social, political and economic events at the local, national and international levels.***

A case in point that has implications for each of these headlines is the oft-quoted misstatement: Americans are living longer today. That statement is incorrect in that the length of time a human being can live has not changed much throughout history. A more correct statement would be: More Americans are living long lives. While both of these developments would have an impact on programs for the elderly, the ultimate consequences of the respective developments would be different. The impact on society of aging is a function of many people living a long time rather than some people living a very long time.

While a good case can be made for the usefulness of applied demography as a means of interpreting and understanding social phenomena, the ultimate goal of any applied science is to effect change. Knowledge gained through the application of demographic concepts, techniques and data has some value in its own right, but the real payoff comes when this knowledge is used to solve a social problem and bring about positive change. In the cases of Medicare and Social Security cited above, demographic knowledge can offer insights into the issues at hand. For example, does the growing Medicare population mean that this government program will eventually go bankrupt as the trend line might suggest? Does the growing number of retirees bode ill for the Social Security program?

Knowledge drawn from health demography helps to shed some light on this as we realize that the major surge in Medicare enrollees for the foreseeable future will be baby-boomers. This generation is living longer than any previous generation of seniors, has more resources to maintain its health status longer, and can generally expect to remain healthy up into their 70s and 80s. While this does not mean that the surge in elderly Americans will not eventually affect Medicare's viability, the situation viewed in this light does not support a "doomsday" scenario.

***By analyzing trends in the demographic behavior of society members, demographers can predict the future characteristics of the population.***

While a number of major trends related to demographics are affecting the society as a whole, there is a personal dimension to this as well. As Weeks (2008) points out, the demographic foundation of our lives is deep and broad.

Although demographers are interested in the characteristics and behavior of groups of people, the demographic attributes of our society affect nearly every aspect of our personal lives in one way or another. In fact, the types of personal decisions that we as individuals make have a cumulative effect on population trends. Some of the decisions that affect our daily lives are:

- The decision to get married (or not) and when
- The decision to have children (or not) and when
- The neighborhood in which we choose to live
- The type of occupation one pursues
- The educational level one aspires to
- The choice of political party one supports
- The lifestyle that one chooses to pursue.

The cumulative effect of decisions such as these made by millions of Americans results in a changing demographic profile. In fact, some demographers make a living projecting the future characteristics of populations based on what is known about that population's current demographic behavior. It would not be unusual, for example, for analysts to use demographic methods to predict how long members of a certain demographic group are going to live and, in fact, determine prospectively what diseases they are likely to die from, or for demographers to predict election results based on the demographic characteristics of likely voters. The bottom line is: Knowing the demographic characteristics of a population opens the door to an understanding of a wide variety of attributes of that population.

## Who Applies Demography?

People in every sector of society employ demographic data, often without being aware of it. Increasingly we hear people speak up the "demographics" of that consumer group or the fact that a certain "demographic" always votes for a certain political party. Even though public expressions about demographics are becoming more common, the widespread use of demographic data and methods is not widely appreciated.

Members of the business community, particularly those involved in marketing, pioneered the use of demographic data in the private sector. The application of demographics to business has become so widespread today that virtually no business decision is made in the corporate boardroom without considering the relevant demographics. Whether the decision involves identifying a target audience for a new product, determining the location for a new store, or designing a sales territory, the demographics of the population under question are a critical piece of the puzzle. This process does not just relate to major corporate decisions but affects us as individuals. The fact that we receive certain catalogues, certain types of junk mail, or certain telephone solicitations reflects the information that marketers have about our demographic attributes.

***Members of the business community, particularly those involved in marketing, pioneered the use of demographic data.***

While the business community was the first to recognize the importance of demographics, this sector is far from the only user of such data. In every aspect of American life demographics have become increasingly important. The allocation of government services depends on an in-depth understanding of the characteristics of the population to be served. Those who aspire to political office begin their campaigns with an assessment of the demographics of their prospective constituents. Those involved in urban planning and community development start with the demographics of the geographic area under consideration. The education system depends on an understanding of the number, location and characteristics of school-aged children, and the services provided by the healthcare system are a direct reflection of the characteristics of the patient population.

***While the business community was the first to apply demographics in the private sector, applications of demography to real-world problems are found in every sphere of society.***

Applied demographers work in virtually every industry, from education to manufacturing to healthcare and at agencies in all levels of government. However, it is noteworthy that most of the people applying demography to concrete problems are *not* demographers. Long before applied demography was recognized as a separate discipline, people in government, business and other fields were regularly using demographics as part of their jobs. It is perhaps a testament to the value of applied demography to find people who are not demographers in virtually every industry employing demographics in their efforts to perform their jobs.

The significance of demography to US industry can be seen in the extent to which those involved in the application of demography have been elevated to roles of importance within the corporate structure. While demographic analysis was at one time relegated to the back room as a low-level technical activity, in today's economy we find those in charge of demographic analysis—albeit usually not demographers per se—seated at the table in the corporate boardroom. It is safe to say that very few business decisions in any industry are made today without considering the demographics underlying the issues and/or the demographic implications of that decision.

In most of these examples, demographics are put to good use—good for the users and good for society. But there are cases where less well-intended individuals and organizations may use demographics for less than noble purposes. Much has been made of racial and ethnic profiling used by law enforcement agencies and championed by those who oppose immigration into this country. Certain groups may be discriminated against due to their demographic characteristics—their race or ethnicity, their poverty level, religion or language. Historically, individuals have been excluded from housing developments, social clubs and occupations due to their sex, race or cultural background. As with any aspect of the human condition, demographics can be used for good or ill.

## Applying Demographic Perspectives and Methods

The demographic perspective involves a way of seeing the world within a demographic context or through a demographic "lens". This unique manner of viewing the world provides a framework for interpreting social phenomena and a means of linking social groups to their environments. This perspective leads to a search for interrelationships between demographic variables and other variables associated with a population.

All of us were born into a game that was already under way making each of us in a sense a "victim" of demographic trends. The rules were written long before we were born, so the type of education we aspire to, the type of job that we are able to obtain, the type of person we are likely to marry, and even the political party we are likely to support, are all reflections of the position within the social structure to which we were born. Indeed, it has now accepted by health demographers that the best predictor of our health status is the ZIP Code in which we were born. While members of society are able to control their demographic futures to a certain extent, we are all constrained by the demographic parameters surrounding our position in society.

***Applying demographics to the analysis of societal phenomena allows demographers to see society from a unique perspective, using a "demographic lens" to view the attributes that characterize any population.***

At the same time we are the "creators" of our demographic circumstances through our behavior, and the decisions that we make contribute to demographic trends. Over the course of the past few decades the behaviors that we have exhibited as a population have had significant implications for the demographic attributes of American society. We have fewer (and later in life) marriages, we are having none or fewer (and later in life) babies, we continue to die from health conditions that are preventable, and we continue to migrate out of the Northeast and Midwest to the South and West. These behaviors on the part of individuals affect the demographic attributes of our society. Ultimately, through our individual behaviors we make our demographic bed and then have to lie in it.

It was noted earlier that demography can be thought of as a "two-way street," and this is manifested through the interaction of population attributes and demographic processes. The composition of a population has implications for the various demographic processes that occur while, at the same time, the current demographic processes have consequences for the population's attributes.

Currently, the US population is exhibiting significant differences in fertility rates as a result of its population composition. The birth rate for the non-Hispanic white population is lower than average while the rate for African Americans, Hispanics and certain other racial and ethnic groups is higher than average. These behaviors attributed to demographically different segments of the population have a significant impact on overall fertility patterns. When these trends are extrapolated into the future, it is predicted that high-fertility minority groups will constitute the majority of the population and non-Hispanic whites will constitute a numerical minority. Numerous other examples of the interdependence of demographic traits and demographic processes can be cited. A conclusion generated by looking at the population through "demographic eyes" is that virtually no demographic or social attribute is randomly distributed within society.

When we look at developments seemingly unrelated to demographic factors, we inevitably find a demographic undercurrent if not a direct causal effect. Examples ripped from the headlines related to such diverse issues as climate change, the international spread of AIDS, an increase in suicide among older Americans, the outsourcing of American jobs overseas, and bankrupt American cities all have underlying demographic components. With the judicious application of demographic concepts, techniques and data, we can develop the demographic perspective required to clearly see the operation of these phenomena.

***No demographic or social attribute is randomly distributed within society.***

One final question to consider as you review the settings for the application of demographic methods is whether or not demographics is destiny. It was noted at the outset that we are born into a game that is already underway and that our demographic circumstances go a long way toward determining our lifestyles and, ultimately, our life chances.

In predicting the future, demographers have advantages over those in many other disciplines for a couple of reasons. First, as indicated above, the interrelationships between demographic variables are such that it is no great leap to be able to project the future characteristics of a population based on known current characteristics. If one knows, for example, the current age and sex distribution for a population, it is possible to project the fertility and mortality patterns for that population ten or twenty years into the future. Of course, the analysts is making certain assumptions about fertility rates and mortality rates and, even more risky, about the behavior of individuals. Even so, such projections are often made and usually with surprising accuracy. If we add other variables, e.g., marital status, to the mix, the ability to predict future fertility levels is enhanced. The predictions could be further refined by profiling the population under study in terms of its racial makeup or socioeconomic status.

***Demographers have an advantage when it comes to predicting the future in that they can anticipate future social change by observing current demographic trends.***

One other way in which demographers have an edge over other researchers is by taking advantage of the natural aging process exhibited by a population. In the absence of high levels of in-migration or out-migration and with some knowledge of reproduction rates, it is possible to project characteristics of a population into the future based on age distribution alone. For example, demographers can project the number of children entering kindergarten in five years because those children are already in the population (i.e., members of the population one year old or less).

Similarly, demographers can project the number of youth entering college ten years from now because they know the number of eight-year-olds currently in the population. Or, demographers can project the number of new "senior citizens" that will be added to the population thirty years from now because they know how many 35-year-olds are currently in the population (as well as the number of those who are likely to die over the next 30 years). Although in each case the demographer has to make certain assumptions (e.g., college admission rates, mortality rates), the natural aging of the population allows the analyst to use "cohort analysis" to project the future characteristics of the population.

***Rather than only observing first-order relationships between variables, the interactions of variables need to be considered.***

One final consideration with regard to the application of demographics to real world problems relates to the interpretation of observed relationships. Throughout this text there have been numerous examples cited of correlations or associations of attributes with different groups within US society. We have ascribed various attributes to different age groups, racial groups, religious groups and so forth. It is tempting to latch on to first-order relationships without considering potential confounding factors, and the danger of doing this lies with the tendency to examine population attributes in isolation from each other.

Perhaps the classic example of this is the long-standing contention that African Americans exhibit worse health status than non-Hispanic whites *because* they are African American. This is not a far-fetched conclusion since on virtually every indicator of health status, African Americans exhibit worse rates. In a now classic study, Williams and Mohammed (2013) demonstrated that, when socioeconomic status was controlled for, virtually all of the presumed racial differences in health status disappeared. Admittedly, there was a residual amount of difference that was attributed to the stress associated with being African American in US society. Even here, this could not be considered an inherent trait of members of this race.

The lesson here is that interrelationship between variables are important and must be identified in order to conduct rigorous analysis of demographic attributes. Just as no attributes are randomly distributed within society, few if any attributes represent the sole contributor to the characteristics of a demographic group. Effective application of demographic methods to the study of society requires diligence in pursuit of attribute interrelationships.

## Settings for Applied Demographics

While all applied demographic activities have certain attributes in common, various specialty areas have evolved sector-specific methodologies. The sections that follow describe some of the common areas within which demographic methods are applied to real-world problems.

Business. Business demography is defined as the application of the content and methods of demography to business challenges and opportunities. Applied to the world of commerce, the demographic perspective involves an understanding of how population processes affect business conditions. Business demographers utilize demographic data, methods, and theory to help decision makers make better choices with respect to commerce-related strategies and the implementation of those strategies.

***The use of demographic data and methods in business planning has a long history.***

The use of demographic data and methods in business planning has a relatively long history, and the early acceptance of demographic input on the part of business professionals contributed to the development of applied demography as a distinct discipline. In fact, the application of demographic resources in the business arena predates its application in most other contexts. As business strategy in general, and marketing in particular, began to move away from a mass market approach to a more targeted strategy focusing on market segments during the 1960s, demographic data become increasingly important.

The growth of business demography was given impetus through: (1) an appreciation on the part of the business community of the role that demographics could play; (2) the increasing interest in business demography on the part of professional demographers; and (3) the increasing ease of access to demographic data. Given this emphasis on advancing society's economic success and the interest of individual businesses, it is not surprising that this was the realm in which demographics was first intensively applied.

Business demographers fall into three distinct groups: those employed by private companies, whose work pertains specifically to those companies and their business activities (e.g., market analyses, customer profiles, site selection); those employed by firms that create demographic databases (e.g., population estimates, consumer spending, lifestyle clusters); and individual consultants who undertake specific projects for particular clients. Box 13.2 describes the use of life cycle analysis in business decision-making.

**Box 13.2: Life Cycle Analysis**

For decades marketers have viewed the consumer life cycle as a framework for guiding the development of marketing plans. The life cycle is a demographically based model that depicts the path that individuals follow as consumers from birth to death. This model contends that individuals experience fairly standardized stages of life as they age, with each stage predictably transitioning to the next stage. Thus, babies are born and spend a certain amount of time in infanthood. Infants become toddlers and then preschoolers. Children experience a period of schooling and in the process become teenagers. Teenagers grow into young adults who then become college students or enter the workforce. Young adults typically marry and start families. They pursue careers and as older adults and eventually become part of the retirement cohort.

The traditional model assumes a set of demographic attributes characterizes each phase of the life cycle. In this model, virtually everyone goes through a period of family formation that includes establishing a household, perhaps buying a home and certainly acquiring the home furnishings required by new householders. As consumers age—and presumably generate more income—their needs and wants change. As their children age their purchase practices change until the last children leave the home. At this point, these aging consumers look to vacation homes, leisure activities and other forms of consumer expenditure to occupy their "golden years." Savvy marketers will be able to determine the consumer needs at each step of the way and promote the goods and service demanded at that point in the life cycle.

It can be argued today that the notion that there is a standard set of phases that virtually everyone goes through is an awkward fit with the realities of life in contemporary American society. This had led to a rethinking of the significance of life cycle analyses. Nevertheless, marketers in the corporate world argue that there is enough predictability when it comes to the consumer's stage in life and their spending activities to support the model's continued use. The challenge is to adapt what has been a useful analytical model to the modern world.

Education. Education planners use demographics to plan for school construction, determine staffing requirements, plan school bus routes, and budget for the provision of educational services. School administrators contemplating the construction of an additional school would use demographics to determine how many current and future school-aged children there are, where they are located, and what their characteristics are (e.g., age, ethnicity). They would also use demographics to examine future birth rates and immigration patterns that might affect the demand for schools.

Educational demography (or the demography of education) involves the application of demographic concepts, techniques and data to issues within the educational arena. The problems that might be addressed by demographers in this sphere include: planning for school construction; determining teacher requirements; determining English-language training requirements; estimating the demand for pre-school programs; and monitoring the racial mix for a school system, among others. Demographic analysis is useful for predicting the demand for educational programs, and the "demographic crisis" being faced by our institutions of higher learning are discussed in Box 13.3.

**Box 13.3: College and the "Demographic Cliff"**

Today's colleges and universities face the prospect of a "demographic cliff" that portends hard times for the foreseeable future. Thanks to lower birth rates during the 2008–2009 recession the college-age population will begin to shrink in 2025. College officials call this demographic phenomenon the "enrollment cliff". The effects will vary based on geographic region and institutional type, but few academic institutions will be able to avoid the effects of this demographic development.

Forecasting the number of college-age youths is a relatively simple task. Take the birthrate for a given year and fast forward 17–18 years—when most kids start college. During the Great Recession with its economic stress and uncertainty, people had fewer children. The number of kids born between 2008 and 2011 plummeted dramatically.

In the graphic below, the precipitous descent resembles a cliff beginning in 2025. Over the succeeding four years, the number of 18-year-olds will decrease by 15%. During that four-year span, colleges will lose approximately 576,000 students. Unfortunately for higher education, this situation is not an aberration. College enrollments have already been declining steadily since 2012 and between 2019 and 2022 undergraduate enrollment dropped by 7%.

It is not just demographics and pandemics affecting college enrollment. A strong job market and low unemployment rates mean companies are willing to pay high wages to attract employees and persuade some high school graduates to put off college temporarily, if not permanently. Add that four-year opportunity cost together with student loan debt and the calculus does not make sense to a growing number of young Americans.

For many college administrators it does appear that "demographics is destiny".

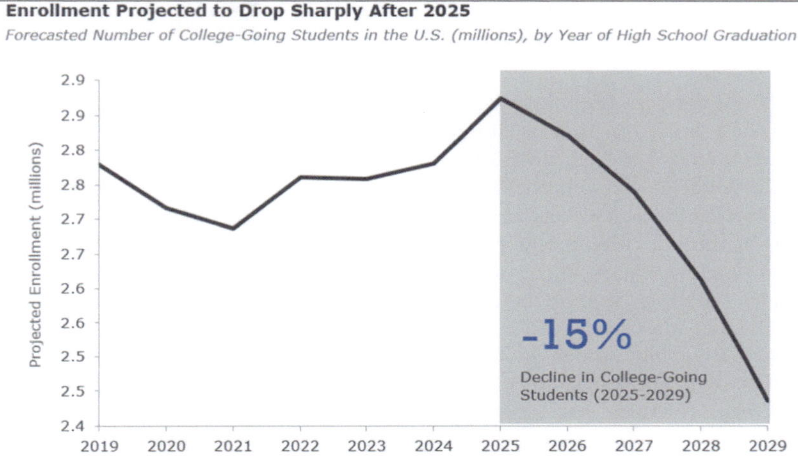


*Source* Drozdowski (2023) and Tasneem (2019)

Politics. Politicians, political consultants and policy makers use demographics to draw up political districts, analyze voting patterns, develop political campaigns and devise advocacy strategies, among other activities. A politician embarking on a political campaign would use demographics to profile the electorate in terms of its demographic attributes and voting patterns, determine the subset of potential voters to target, and develop a demographically based promotional strategy targeting potential voters. As noted earlier, demographic considerations underlie many of the programs formulated by policy makers.

Applications of demographic concepts, methods and data have principally occurred in two areas: the electoral process and the development of public policies. In the realm of politics the implications of demographics are ubiquitous. While there is some overlap between politics and policy, most consider them distinct realms of endeavor with their respective goals and practitioners. Box 13.4 describes the implications of demographic trends for the future of American politics.

**Box 13.4: Demographic Change and the Future of the Republican Party**
It is often said that "demographics is destiny," and that may be the case with American politics. Voters who support different parties or exhibit different political orientations (e.g., conservative, liberal, libertarian) have historically exhibited different demographic characteristics. Although there have been major shifts in political orientation by demographic groups (e.g., the shift of Southern white voters from Democrat to Republican after civil rights legislation was passed), the demographic characteristics exhibited by voters that fall along party and philosophical lines have remained consistent over time. In fact, since 2000 these lines have become increasingly rigid as both major political parties (and

the Republican Party in particular) have discouraged any dissension from the "party line".

Given the demographic profiles of those supporting different political parties and/or political orientations, it might be worthwhile to examine current demographic trends and their implications for the future electorate. Over the past 15 or 20 years the Republican Party has become the party of old, white men for all practical purposes. This was the only group that broke heavily for Republicans in the last three presidential elections. This support came primarily from working-class whites, those with *some* college education, and upper-income groups (although support here appears to have eroded).

On the other hand, those supporting the Democratic party in recent years tended to exhibit much more diverse demographic traits. These supporters include women (particularly single women), younger voters, and the well-educated. However, the most significant trend is the increasingly one-sided support for Democrats on the part of racial and ethnic minority groups.

African Americans were totally on board for the Democratic candidates in recent presidential elections. Hispanics, Asian-Americans, and Jewish voters are reliably Democrat in their orientation. The bottom line? The segments of the population that now support the Republican Party are declining in number while the groups that are increasingly throwing their support behind the Democrats are increasing in number.

In speculating about the implications of demographic trends for the future political process certain assumptions must be made. First, it is assumed that the native-born non-Hispanic white population will continue to decline as a proportion of the population. It is also assumed that virtually all racial and ethnic groups (other than non-Hispanic whites) will continue to increase in number and proportion for the foreseeable future. It is also assumed that non-white racial and ethnic minority groups will continue to reproduce at a higher rate than the non-Hispanic white population. Even under the most conservative assumptions, many demographers argue, the proportion of the electorate supporting the Republican Party will steadily decline, while the proportion of the electorate supporting the Democratic Party will steadily increase.

Labor force development. Policy makers, government officials and education planners use demographics to calculate unemployment rates, monitor the growth/decline of various industries, predict the future demand for specific occupations, and project the future worker pool. A job training program would use demographics to profile the local workforce, determine employment/unemployment levels, identify the types of jobs needed in the local economy, and recommend job training initiatives that would equip the workforce to fill the available jobs.

Any analysis of labor force trends and the workers that participate in it must have a strong underpinning. Past trends in labor force participation tell us a lot about the changing nature of both work and the labor force itself. The attributes of the future labor force will be extensions of today's demographic characteristics. Based on current demographic knowledge, we know that the labor force is getting older (with fewer young people to replace them), becoming more racially and ethnically diverse, and including more females than males in many occupations.

Demographers examine trends in industry and the patterns of employment that this involves. They couple this with the future characteristics of the workforce to determine the future configuration of US industry. Box 13.5 lists the top 10 fastest growing jobs in the future.

**Box 13.5: Fastest Growing US Occupations 2021–2031**

| *Occupation* | *Percent increase* |
|---|---|
| Nurse practitioners | 46 |
| Wind turbine technician | 44 |
| Theatre ushers/other employees | 41 |
| Motion picture projectionists | 40 |
| Restaurant cooks | 37 |
| Data scientists | 36 |
| Athletes/sports competitors | 36 |
| Information security analysts | 35 |
| Statisticians | 33 |
| Sports officials | 32 |
| Web developers | 30 |
| Animal caretakers | 30 |
| Choreographers | 30 |
| Taxi drivers | 28 |
| Health services managers | 28 |
| Logisticians | 28 |
| Physician assistants | 27 |
| Solar energy installers | 27 |
| Animal trainers | 27 |
| Physical therapy assistants | 26 |

*Source* Bureau of Labor Statistics (2022)

Transportation. Transportation planners use demographics to predict the demand for roads/highways, plan for public transportation, perform cost–benefit analyses for new roads, and assess the impact of transportation projects on the community. The design of our transportation system with its emphasis on automobiles (at the expense of public transit and other forms of travel) has been and continues to be a major force in the determination of our social structure. For a century, the need for automobile transportation has served to reconfigure our social environment.

A transportation planner conducting a study of existing public transportation to determine needed modifications would use demographics to profile the population currently using public transportation, identify additional potential users, determine current and future destinations for riders, and recommend changes in routes and schedules.

Because Americans are a very mobile population, transportation planning is of paramount importance although often neglected. We spend an inordinate amount of time in our cars, and access to automobiles is a requirement for effective living for most of the US population. Box 13.6 discusses the impact of highway design on our communities.

**Box 13.6: Highway Design and Urban Development**

When Congress approved the Federal-Aid Highway Act of 1956, it authorized what was then the largest public works program in US history. The law promised to construct 41,000 miles of an ambitious interstate highway system that would crisscross the nation, But the highway expansion, implemented largely between the late 1950s and the early 1970s, came at a huge cost to America's urban communities of color.

According to estimates from the US Department of Transportation, more than 475,000 households and more than a million people have been displaced nationwide because of federal roadway construction. Hulking highways cut through neighborhoods, darkened and disrupted the pedestrian landscape, worsened air quality and torpedoed property values. Communities lost churches, green space and whole swaths of homes. They also lost small businesses that provided jobs and kept money circulating locally—crucial middle-class footholds in areas already struggling from racist zoning policies, disinvestment and white flight.

The neighborhoods destroyed and families uprooted by highway projects were largely black and poor and that was by design. Highway construction was considered a convenient way to raze neighborhoods considered undesirable or blighted. Hardly a major city with a significant minority population went unscathed by the legislation: New York, Miami, Chicago, Minneapolis, Pittsburgh, Oakland, Nashville, Baltimore, Atlanta—and more.

By the time the interstate highway system was completed, it had fundamentally restructured urban America. Many believed that the best way to eradicate slums where Black people lived was to build highways through them. For those who wanted to flee the encroachment of blacks altogether, the new Interstates only helped speed white flight from the city.

*Source* Evans, F. (2021). "How Interstate Highways Gutted Communities—and Reinforced Segregation," *History.com*. Downloaded from URL: https://www.history.com/news/interstate-highway-system-infrastructure-construction-segregation

Community Development. Community planners use demographics to predict the future demand for various types of housing, develop land use plans, plan public housing projects, project the level of homelessness, and determine the supply of affordable housing. A community development corporation might use demographics to assess the level of home ownership in the community, track housing sales, monitor foreclosure trends, and assess the mobility of community residents with the objective of encouraging housing stability within the community.

Demographers often play a role in community development projects. Since such projects generally affect the lives of people, it is often necessary to profile the population to be affected and assess its attributes. One type of assessment that a demographer may be involved in is an environmental impact statement (EIS). The nature of the EIS is described in Box 13.7.

**Box 13.7: Environmental Impact Statements**
The environmental impact statement (EIS) is a government document that outlines the impact of a proposed project on its surrounding environment. In the United States, these statements are mandated by federal law for certain projects. Environmental impact statements are meant to inform the work and decisions of policymakers and community leaders.

An EIS is a report mandated by the National Environmental Policy Act of 1969 (NEPA) to assess the potential impact of actions "significantly affecting the quality of the human environment." This requirement under NEPA does not prohibit harm to the environment, but rather requires advanced identification and disclosure of potential harm. Development projects that constitute major federal action, as defined by law, including those that use federal land, federal tax dollars, or are under federal agency jurisdiction, are required to assess the impact of a proposed project on the physical, cultural, and human environments affected by the proposed project.

An EIS outlines the status of the environment in the affected area, provides a baseline for understanding the potential consequences of the proposed project, identifies positive and negative effects for the environment, and offers alternative actions, including inaction, in relation to the proposed project.

Environmental impact statements often address local areas and projects that are tangible and potentially well-known, making them widely available and accessible. They are tools for informing the public about the development and engineering of the built environment, as well as windows into the civic planning that shapes communities. They reveal details about the environmental health and development of a specific area in a comprehensive report.

The EIS analysis does not just assess the natural environment but considers the social environment and the project's impact on the lives and health of affected residents. This is where a demographer may have input, reviewing the implications of the planned project or policy change on the lives of community residents. This would involve a detailed assessment of those affected and the manner in which the project would affect them. Some of these impacts would be direct (such as the introduction of pollutants to an area); others would be more indirect (such as the implications of new roadways and an increase in neighborhood traffic). A project may have implications for educational opportunities, job options and transportation resources, all of which have the potential to alter the lives of those affected by the project.

Social Services. Social service agencies use demographics to identify vulnerable populations and profile demographically defined groups in need of social services, locate social service offices, develop welfare policies, and evaluate the effectiveness of social service programs. A government agency might use demographics to develop welfare policies, identify the in-need population, determine this population's characteristics, examine the factors contributing to neediness, and structure assistance programs to address these needs.

The factors that determine the need for social services are the types of variables that demographers deal with on a regular basis. The level of need and the type of services needed are typically framed in terms very familiar to demographers. A social services assessment on a client would involve a review of household structure, educational level, employment status, income level (especially if poverty is a factor), housing conditions, access to healthy food, access to transportation, and language proficiency.

While a demographer is not likely to deal with individual clients, he or she would be able to profile the community from which the client is drawn in terms of the demographic attributes that may have implications for his or her needs. It has become increasingly important to consider the social context of those in need of services in order to develop an understanding of their situation. This type of community profile is the bread-and-butter of demographic analysis.

Healthcare. Health planners and administrators use demographics to identify medical service areas, determine the demand for health services, select locations for health facilities, measure disparities in health status and health services utilization, and develop marketing plans to promote health services and products. A health planner would use demographics to determine the size and characteristics of the population in a health service area, the types of health problems likely to affect this population, and the types of personnel, services and facilities required.

In recent years the subdiscipline of health demography has emerged in recognition of the importance of the social dimensions of health and healthcare. It has become increasingly apparent that our population's health conditions are more a reflection of social conditions than they are of pathogens in the environment.

The social determinants of health and illness are being increasingly recognized and these are the variables that demographers deal with on a daily basis. The factors that contribute to poor health within a population include poverty, poor housing, lack of grocery stores, lack of jobs and so forth.

When the COVID-19 pandemic emerged in 2020, health demographers were called upon to make sense of the factors surrounding the disease and to predict the impact of the pandemic on various sectors of the population based on their demographic traits. Box 13.8 examines some of the demographic implications of the pandemic.

**Box 13.8: Demographic Implications of the COVID-19 Pandemic**
When the COVID-19 pandemic struck the American population, no one knew much about its characteristics or whom it would affect. Even before government bodies at any level could react, demographers had begun analyzing the available data and developing models for predicting the course of the pandemic.

As with virtually every disease, the impact is different for various segments of the population, and the COVID-19 pandemic was no exception. When demographers examined the data that were becoming available, they found that the epidemiological data revealed that, in fact, certain segments of the population were more affected than others. The first "reveal" was that seniors were among the earliest to be infection and recorded relatively high death rates. A second category that demographers identified was members of certain racial and ethnic groups. These included African Americans, Hispanics and American Indians. The risk was increased for those who were among these groups but also elderly.

A third category of infected parties were those who worked in essential occupations that carried a high risk of exposure. While most white-collar workers could move to the safety of home, many others had to continue working in crowded and unsanitary conditions. It did not help that many of these at-risk people were also members of high-risk minority groups. They had the additional risk of living in crowded, unsanitary conditions with little opportunity for social distancing.

An interesting twist to this epidemiological review was introduced by virtue of the behavior of US citizens in the face of the pandemic. Once well into the pandemic, the data revealed some interesting developments related to the distribution of disease within society. Members of minority populations increased their level of vaccination and other harm-reduction activities, while members of the white population lagged behind. Those who had initially fared well in the face of the pandemic began exhibiting higher rates of infection and death than those who were initially over-represented.

This led to the identification of another attribute that contributed to high incidence rates: political ideology. It was found that those who identified as conservative Republicans were less likely to be vaccinated and, thus, more likely to get sick and even die than those who identified as Democrats. Because of the careful research carried out by applied demographers, it was possible to more accurately determine the nature of the disease and its impact on the US population.

Criminal Justice. Law enforcement officials use demographics to identify the characteristics of criminals and victims, identify high-crime areas, plan for correctional facilities, determine deployment patterns for police, and develop community crime prevention programs. Law enforcement officials may use demographics to determine priorities for enforcement, identify the types of crimes that are the most serious and/or prevalent, determine the spatial distribution of these crimes, calculate the cost–benefit of focusing on certain crimes rather than others, and predict the consequences of various law enforcement initiatives.

When demographers have examined the demographic correlates associated with our criminal justice system, they have found significant disparities in terms of the operation of the system. These disparities are based primarily on race and ethnicity. A review of the data indicates that, although whites and Blacks commit crimes at essentially the same rate, blacks are much more likely to be arrested than whites. Blacks who are arrested are more likely to be convicted than whites charged with the same crime. Of those who are found guilty, Blacks on the average are given harsher sentences. As a result of the disparities in the operation of the system, a significant proportion of the non-white population is incarcerated or has a criminal record. Further evidence of the operation of the system is the fact that minorities make up one-fourth of the population but account for almost two-thirds of the prison population.

## Key Points

- Applied demography involves the application of demographic theories, concepts, methods and data to the solution of "real world" problems.
- Having access to information allows demographers to contribute to policy making.
- Applied demography provides a means of interpreting social, political and economic events at the local, national and international levels.
- By analyzing trends in the demographic behavior of society members, demographers can predict the future characteristics of the population.
- Members of the business community pioneered the use of demographic data within the private sector.
- While the business community was the first to apply demographics, applications of demography to real-world problems are found in every sphere of society.
- Applying demographics to the analysis of social phenomena allows demographers to see society through a "demographic lens".
- Demographers have an advantage when it comes to predicting the future in that they can anticipate social change by observing current demographic trends.
- Rather than only observing first-order relationships between variables, the interactions of variables need to be considered.

## Discussion Questions

13.1 Why do you think demographers are well positioned to analyze social issues within our society? What information and methods do they have that make them particularly qualified to conduct these analyses?

13.2 What are some of the disparities in our society that demographers have identified? Think of examples in such diverse areas as educational achievement, income, health status, and criminal justice. Do you see ways in which these variables may interface with each other?

## Exercises

*Exercise 13.1: The Consequences of Current Demographic Trends*

A number of demographic trends that are currently affecting the US population are presented below. Review each of the trends and discuss the implications of these trends for our healthcare system. Consider the ways in which these trends are affecting or will affect the types of health problems that we can expect, the types of services that will be utilized and any other consequence related to healthcare that might result from these trends. Be prepared to explain your conclusions to the rest of the class.

- The US population is rapidly aging as the baby boom cohort moves into its senior years, the elder population continues to grow, and the working age and young adult populations continue to decline.
- The US population is becoming more "feminized" as the population ages and women constitute an ever-growing proportion of the population.
- The US population is becoming increasingly diverse in terms of race and ethnicity, with the Hispanic population growing dramatically and significant increases in a number of other ethnic groups.
- The "traditional" family is becoming less common in US society as fewer people marry and childless couples, single-parent households, cohabiting (sometimes gay) couples, and individuals living alone become more common.
- The US population is becoming more diverse in terms of religion but less religious overall in terms of formal religious participation.
- The income gap between the "haves" and the "have nots" in the US is increasing and the poverty population has begun to grow after experiencing more than two decades of decline.

*Exercise 13.2: Projecting the Demand for Educational Services*

Demographers are often called upon to generate estimates of the demand for educational services. They may be asked for figures on the number of children expected to enter kindergarten at some point in the future, or the number of expected to enter high school or college. A useful technique for generating such estimates is cohort analysis, a method that allows demographers to take advantage of existing information. Based on the table showing the youth population below, students should project the number of children who will enter the first grade (age 6) in five years, the number of teens who will enter high school (age 15) in eight years, and the number of students who will be eligible for college (age 18) in ten years.

| *Age (years)* | *Number* |
|---|---|
| 1 | 1174 |
| 2 | 910 |
| 3 | 766 |
| 4 | 886 |
| 5 | 2515 |
| 6 | 2659 |
| 7 | 2300 |
| 8 | 1629 |
| 9 | 1581 |
| 10 | 1413 |
| 11 | 1557 |
| 12 | 1988 |
| 13 | 1605 |
| 14 | 1149 |
| 15 | 1671 |
| 16 | 1479 |
| 17 | 1359 |
| 18 | 1311 |

Provide the following figures: Number entering the first grade in 5 years: ____
Number entering high school in 8 years: ____
Number eligible for college in 10 years: ____

## References

Bureau of Labor Statistics. (2022). Fastest growing occupations. Occupational Outlook Handbook. https://bls.gov/ooh/fastest-growing.htm

Drozdowski, M. J. (2023). Looming enrollment cliff poses serious threat to colleges. Best Colleges. https://www.bestcolleges.com/news/analysis/looming-enrollment-cliff-poses-serious-threat-to-colleges/#:~:text=What%20Is%20the%20Enrollment%20Cliff,when%20most%20kids%20start%20college

Rives, N., & Serow, W. (1984). *Introduction to applied demography: Data sources and estimation techniques*. Sage Publications.

Tasneem, A. (2019). What the looming demographic storm means for your state. https://eab.com/insights/expert-insight/enrollment/what-the-looming-demographic-storm-means-for-your-state/

Weeks, J. R. (2008). *Population: An introduction to concepts and issues*. Thompson Wadsworth.

Williams, D. R., & Mohammed, S. A. (2013). Racism and health I: Pathways and scientific evidence. *American Behavioral Science, 57*(8), 1152–1173.

## Additional Resources

Martins, J., Yusuf, F., & Swanson, D. (2011). *An introduction to consumer demographics and behaviour: Markets are people*. Springer B.V. Press.
Meade, M., & Emch, M. (2010). *Medical geography* (3rd ed.). Guilford Press.
Murdock, S. H., & Swanson, D. (2008). *Applied demography in the 21st century: Selected papers from the biennial conference on applied demography*. Springer.
Shyrock, H. S. (2013). *The methods and materials of demography* (2nd ed.). Academic Press.
Swanson, D. (Ed.). (2016). *Frontiers of applied demography*. Springer.
Swanson, D., & Verdugo, R. (Eds.). (2023). *Socio-demographic perspectives on the COVID-19 pandemic*. Information Age Publishing.

# Glossary

**Abortion** The deliberate termination of a human pregnancy.

**Activities of Daily Living (ADL)** The tasks that individuals must perform in order to take care of themselves (e.g., dressing, toilet use), used to assess the level of disability.

**Acute condition/disease** A health condition characterized by rapid onset, usually short duration, and a clear-cut disposition (e.g., cure, death), typical of developing countries and younger populations.

**Administrative records** Data collected routinely during the course of operations by organizations and government agencies; certain national registries are based on administrative records.

**Age** Age is reported as age at last birthday (i.e., age in completed years), often calculated by subtracting the date of birth from the reference date.

**Age adjustment** A technique used to standardize data in order to compare risks for two or more populations at one point in time or for one population at two or more points in time, thereby eliminating differences in observed rates that result from age differences in population composition.

**Age-adjusted mortality rate** The number of deaths per 100,000 population adjusted to reflect the age structure of the population, thereby accounting for differences in the size of various age cohorts.

**Age-dependency ratio** For a population, the ratio of the number of persons defined as dependent (under 15 years and over 64 years) to the number of persons defined as economically productive (15–64 years).

**Age-sex structure** The composition of a population as determined by the number or proportion of males and females in each age category.

**Age-specific birth rate** The number of births per 1000 women within a particular age cohort for a particular time period.

**Age-specific rates** The level of occurrence (per 1000, 10,000 or 100,000 population) of a phenomenon for specific age cohorts.



R. K. Thomas, *Demography: An Introduction to Population Studies*, Springer Texts in Social Sciences, https://doi.org/10.1007/978-3-031-56623-3

**Aging of population** A process in which the proportions of adults and elderly in a population increase, while the proportions of children and adolescents decrease.

**American Community Survey (ACS)** A national sample survey initiated in 2005 for annually collecting data that were historically obtained through the long form administered during the decennial census.

**Area of Dominant Influence (ADI)** The geographic territory covered by a particular form of media (e.g., radio, television).

**Assumption** A fact or situation that is accepted as true and used as a basis for positing a theory or projecting change.

**Average annual rate of percentage change** A method for calculating estimates and projections based on interpolating or extrapolating change based on a known percentage change from period to period.

**Baby Boom** A time period in the US exhibiting a dramatic increase in fertility rates and in the absolute number of births; generally considered the period between 1946 and 1964.

**Baby Boomers** The segment of the US population (born between 1946 and 1964) and representing the largest cohort in the age distribution in the US up to that point in time.

**Baby Bust** A period in US history characterized by a rapid decline to record low fertility rates during the period immediately following the baby boom.

**Balancing equation** A basic demographic formula used to estimate total population change between two points in time—or to estimate any unknown component of population change, provided that the other components are known.

**Biosocial variable** A demographic attribute reflecting a biological trait that has a significant social dimension (e.g., age, sex and race).

**Birth certificate** The standard form completed for each birth that takes place in the US, typically collected by local health departments and ultimately submitted to the National Center for Health Statistics.

**Birth cohort** A birth cohort consists of all persons born within a given period of time, such as a calendar year.

**Birth rate** A measure of the level of fertility for a population calculated by dividing the number of live births in a population during a year by the resident population.

**Birth weight** The weight of a newborn obtained at the time of birth.

**Carrying capacity** The maximum sustainable size of a resident population in a given ecosystem.

**Case analysis** A technique used by epidemiologists to identify the incidence of specific health conditions within a population and track the progression of these conditions over time.

**Case fatality rate** The proportion of persons contracting a disease who die from it during a specified time period.

**Case rate** The number of reported cases of a specific disease per 100,000 population in a given year.

**Cause of death** The reason for the death of an individual entered on the standard death certificate and used in mortality analyses.

**Cause-specific mortality rate** The death rates for a population attributed to different causes of death, typically stated as a rate per 100,000 for a specified year.

**Center for Medicare and Medicaid Services (CMS)** The federal agency with responsibility to managing the Medicare and Medicaid programs.

**Census** A data collection technique that involves obtaining data from the entire population.

**Census block group** The aggregation of census blocks into groups for data analysis purposes; the lowest level of geography for which the American Community Survey provides data.

**Census block** The smallest geographic area for which the Census Bureau collects and reports data during the decennial census; a square or other polygon that is formed by the streets that comprise the four (usually) sides of a block.

**Census Bureau, US** The federal agency within the U.S. Department of Commerce with responsibility for conducting the decennial census and a variety of other censuses and survey activities.

**Census tract** A geographic area established by the Census Bureau for the collection and reporting of census data; typically includes between 2000 and 4000 residents.

**Centers for Disease Control and Prevention (CDC)** The primary public health agency within the federal government with responsibility for monitoring health problems and addressing identified issues.

**Central city** The core city of a metropolitan statistical area (typically with a minimum of 50,000 residents) or within a micropolitan statistical area (typically with 10,000–50,000 residents).

**Central county** The core county within a metropolitan statistical area, typically including the central city.

**Change, population** A general term for any change over time in a population's size or characteristics.

**Childbearing years** The reproductive age span of women, assumed for statistical purposes to be 15–44 or 15–49 years of age.

**Child-woman ratio** The number of children under age 5 per 1000 women of child-bearing age in a population in a given year.

**Chronic disease** A health condition that involves gradual onset and a permanent state that can only be managed and not cured.

**Closed population** A population with no migratory flow either in or out, so that changes in population size occur only through births and deaths.

**Cohort** A group of people sharing a common temporal demographic experience who are observed through time, often based on age but could reflect a common experience (e.g., people graduating from college in a particular year).

**Cohort analysis** Observation and measurement of a cohort's demographic behavior over time.

**Cohort change ratio technique** A technique for estimating or projecting population statistics that involves advancing an existing age cohort over time and using a child-women ratio to supplement the advancing cohorts.

**Cohort effect** The observed impact that certain experiences and conditions exert on a cohort of individuals as that cohort is monitored over time.

**Collection, data** Any one of a number of procedures (both primary and secondary) for acquiring the data required for the implementation of a research plan.

**Commercial data vendor** A private organization that collects, compiles, analyzes, and disseminates data.

**Communicable (or infectious) disease** A disease caused by a microorganism that can be spread directly or indirectly from one person to another.

**Community survey** A type of sample survey conducted on the community at large, typically using households within the community as the sampling frame.

**Completed fertility rate** The number of children born per woman to a cohort of women by the end of their childbearing years.

**Composition, population** The demographic characteristics exhibited by a population.

**Component method** A form for synthetic data generation that involves determining the combined impact of changes in fertility, mortality and migration.

**Concentration, population** An assessment to the manner in which population is distributed over a specified territory; the extent to which a population is grouped together or dispersed.

**Constant change method** A population estimation/projection method that examines population change trends for two or more points in time and generates figures that reflect either annual average percentage change or annual average percentage change.

**Contraception** Any method used to prevent the fertilization of a human egg.

**Crude rate** Rate of any demographic event computed using the entire population as the denominator in the equation.

**Crude Birth Rate (CBR)** A measure of the level of fertility of a population based on the number of births per 1000 population; also referred to as the crude fertility rate.

**Crude Death Rate (CDR)** A simple measure of the level of mortality of a population based on the number of deaths per 1000 population; the crude death rate may be misleading since there are significant variations in death rate by age. Also referred to as the crude mortality rate.

**Data standardization** A technique by which rates are recalculated using a "standard" population in order to adjust for an abnormal age structure or other anomaly in demographic traits.

**Death certificate** The standard form completed for every death that occurs in the US, typically collected by a local health department and ultimately submitted to the National Center for Health Statistics.

**Death rate** The number of deaths per 1000 population in a given year.

**Demand (for health services)** The combined healthcare needs and wants of a population that constitutes the volume and type of health services "demanded" by that population.

**Demographic transition** The historical shift of birth and death rates from high to low levels in a population. The decline of mortality usually precedes the decline in fertility, thus resulting in rapid population growth during the transition period.

**Demography** The scientific study of human populations, including their sizes, compositions, distributions, densities, growth, and other characteristics, as well as the causes and consequences of changes in these factors.

**Demography, applied** A subdiscipline within demography that emphasizes the application of demographic concepts, methods and data to concrete, real-world problems.

**Density, population** An indicator of the density or sparseness of the population in a geographic area, usually measured in the US in terms of persons per square mile.

**Dependency ratio** An index that compares the number of "dependent" individuals within a population (i.e., children and seniors) to the number of independent individuals (i.e., working-age individuals) that are expected to support them.

**Depopulation** A situation in which a population is experiencing a decline in number.

**Descriptive research** Research that describes the population, market, or situation under study, but does not attempt to explain the patterns that are identified.

**Disability** A condition (physical or mental) of individuals that limits their ability to carry out the activities of daily living.

**Disability rate** The level of incapacity within a population as measured by the number of cases of physically and/or mentally handicapped individuals, or by the reported level of restricted activity (e.g., school-loss days or work-loss days).

**Disease** A term used in a number of different ways but generally referring to a state of pathology within an individual organism.

**Disease of civilization** A term applied to chronic conditions that have arisen as a result of the modernization of society and reflect the impact of environmental factors and lifestyle on health status.

**Disease of despair** A health condition that arises as a result of social or economic conditions of modern society and reflect a state of "despair" on the part of affected individuals.

**Divorce rate** The number of divorces per 1000 population in a given year.

**Econometric model** A projection technique that generates forecasts based on a "best fit" model utilizing a wide range of variables that are thought contribute to future conditions.

**Economic census** An inventory of economic activity conducted every five years by the Census Bureau that compiles certain data for every business in the US; the primary source of data on healthcare establishments.

**Education** The average level of education attained by a population, usually measured in terms of grades or degrees completed and sometimes presented in terms of years completed (attainment). Also measured in terms of a population's current enrollment in educational programs (participation).

**Elasticity** The tendency for the level of a particular indicator to rise or fall in response to social, economic and political factors.

**Emigration** The process of leaving one country to take up permanent or semi-permanent residence in another.

**Emigration rate** The number of emigrants departing an area of origin per 1000 population for that area in a given year.

**Environmental Impact Statement (EIS)** An assessment of the impact of a particular project or policy on the affected environment (with environment seen in the widest possible terms). Typically required by the federal government prior to the initiation of an project with implications for the environment.

**Epidemiological analysis** An approach to the study of health phenomena that involves the relationship between individuals and their social environment and the distribution of health problems among various segments of the population.

**Epidemiologic transition** A change in a population's epidemiologic profile—from acute to chronic health problems—as a result of aging, changing demographic characteristics and lifestyles.

**Estimate** The calculation of a figure for a current or past period using a statistical method.

**Estimate, population** A calculated estimation of the population for a particular area or population category for some current or past time period; a form of synthetic data.

**Ethnicity** A common racial, national, tribal, religious, linguistic, or cultural trait or background characterizing members of a population.

**Etiology** In epidemiology, the cause or causes of an identified disease.

**Exploratory research** Preliminary (often informal) analysis that examines the general nature of the problem or opportunity under study and provides guidance for more advanced analyses.

**Extrapolation** A projection technique that graphically, mathematically or statistically identifies a pattern for a phenomenon and then extends that pattern into the future to predict utilization.

**Exurb** A residential community that exists outside the boundaries of a metropolitan statistical area with important social or economic connections to the central city.

**Family** Two or more persons living together and related by birth, marriage, or adoption.

**Family income** The amount of income generated by family households for a specified time period; total income to families divided by the number of families.

**Fecundity** The physiological capacity of a woman to produce a child.

**Fertility** The actual reproductive performance of an individual, a couple, a group, or a population.

**Fertility rate** One of several metrics developed for calculating the reproductive experience of a population.

**Fetal death** A death that occurs to a fetus between 20 weeks of gestation and nine months of gestation.

**Fetal mortality rate** The number of deaths to fetuses with gestation between 20 weeks and nine months for a specified time period divided by the sum of live births plus fetal deaths for that year per 1000 population.

**Focus group** A qualitative method of data collection that involves obtaining feedback from individuals in a group setting; typically involves 8–10 persons and a trained moderator following an interview guide.

**Forecast** A form of projection that incorporates likely future developments into the calculations.

**Gap analysis** An analytical technique used to assess the gap between the needs of a population and the resources required to meet those needs, or between the current status of an organization and the position in which it would like to be.

**Gender** The social dimension of sex that involves sex roles that exist independent of the physical attributes of the two sexes.

**General fertility rate** The number of live births per 1000 women ages 15–44 (or 15–49 years) during a given year.

**Geographic Information System (GIS)** A computerized application for performing spatial analysis and generating maps reflecting the analysis.

**Geographic unit** A physical area demarcated by defined boundaries and used as a basis for data collection.

**Gestation** The process of producing a birth with the gestation period counted from the first day of the last normal menstrual cycle.

**Globalization** The worldwide expansion and interconnectedness of organizations and their associated economies and influence.

**Gross Reproduction Rate (GRR)** The average number of daughters that would be born alive to a woman (or group of women) during her lifetime if she passed through her childbearing years conforming to the age-specific fertility rates of a given year.

**Group quarters** Any form of congregate living in which unrelated individuals share common quarters and are generally under the supervision of the administrators of the facility (including college dormitories, military barracks, prisons and assisted living facilities).

**Growth rate** The number of people added to (or subtracted from) a population in a year due to natural increase and net migration expressed as a percentage of the population at the beginning of the time period.

**Health** A state of wellness whose definition depends on one's perspective. In the narrowest sense, health has been defined as the absence of disease and disability. In its broadest sense, it has been defined as a state of complete physical, social, mental and spiritual well-being.

**Health status** The overall condition of an individual or population in terms of its health; the level of morbidity characterizing a population.

**Health status indicator** A measure of the relative health condition of a person or population, usually in the form of an index score and generated through self-reports and/or from official statistics produced on the population in question.

**Health/healthcare system** A multifacility healthcare organization; also may refer to the overall healthcare system.

**Healthcare** Any informal or formal activity intended to restore, maintain or enhance the health status of individuals or populations.

**Household** One or more persons occupying a housing unit.

**Housing unit** Any non-congregate living unit to include houses, apartments, condominiums, mobiles homes, and boats.

**Illegal (or undocumented) alien** A foreign national who has entered a country without proper documents or has violated the terms of legal admission to the country.

**Immigrant** An individual is designated as an immigrant by virtue of moving from one country to another with the intention of permanent residency.

**Immigrant paradox** The situation in which immigrants arriving to the US exhibit better health than their countrymen already in the US. The health status of second- and third-generation immigrants tends to decline with longer residence in the US.

**Immigration** The process of entering one country from another to take up permanent or semi-permanent residence.

**Immigration rate** The number of immigrants arriving at a destination per 1000 population at that destination in a given year.

**Incidence** The number of new cases of a disease, disability, or other phenomenon recorded during a specified period of time, reported in terms of the number of cases per 100, 1000, 10,000 or 100,000 population.

**Incidence rate** The number of occurrences of a particular phenomenon per 1000 population at risk, for a given period of time.

**Index** A composite "score" usually calculated from responses, on a survey or the characteristics of a population that have been identified; the index score can then be compared to other respondents or populations.

**Infant death** A death recorded for an infant born alive who dies within the first year of life.

**Infant mortality rate** The number of deaths of infants under age 1 per 1000 live births in a given year.

**Information** Data that have been rendered meaningful by means of statistical analysis and interpretation.

**Intercensal population** Population counts or estimates generated between decennial censuses.

**Interpolation** A technique for generating population estimates that involves calculating a figure between two population data points.

**In-migration** The process of entering one administrative subdivision of a country (such as a province or state) from another subdivision to take up residence.

**Less Developed Country (LDC)** Refers to countries in Africa, Asia (except Japan), Latin America and the Caribbean, and Oceania (except Australia and New Zealand) that are less developed than modern, industrial countries.

**Life cycle** The maturation of a population, a product, or an industry from birth to death.

**Life cycle analysis** An assessment of a population on the basis of its members life cycle configuration.

**Life expectancy** The length of life that can be anticipated for members of a population; the average number of years calculated based on the age of death for all population members.

**Life span** The maximum age that human beings could reach under optimum conditions.

**Life table** A tabular display of life expectancy and the probability of dying at each age (or age group) for a given population, according to the age-specific death rates prevailing at that time.

**Lifestyle** The set of attitudes, values, preferences and behavior patterns that distinguish subsets of the population from each other; often used interchangeably with "psychographic characteristics".

**Lifestyle analysis** An analytical approach that evaluates a population on the basis of the lifestyle characteristics associated with various subsets within the population; often used interchangeably with "psychographic analysis".

**Limitation of activity** A measure of disability that involves a determination of the ability of an individual to carry out various activities. Usually presented in terms of bed-restricted days, school-loss days, work-loss days, etc.

**Living arrangement** The nature of relationships between members of a household; includes such categories as married without children, married with children, living alone, cohabitative relationships, and unrelated individuals living together.

**Malthusian theory** A theory of population growth posited by nineteenth century British economist Thomas Robert Malthus that argued that population would increase until the carrying capacity of the environment was reached, after which war, disease and famine would serve to check population growth.

**Marital status** The official status of individuals in terms of marriage, typically including the categories of never married, married, divorced, widowed and, sometimes, separated.

**Market area** Geographic area considered to be served by a particular organization or company; typically referred to as a "service area" by not-for-profit organizations.

**Marriage rate (or crude marriage rate)** The number of marriages per 1000 population in a given year.

**Maternal mortality** Deaths to women that are associated with child-birth, occurring before the birth, during the birth or after the birth.

**Maternal mortality rate** The number of women who die as a result of pregnancy and childbirth complications per 100,000 live births in a given year.

**Mean age** The mathematical average age of all the members of a population.

**Measurement** The assignment of meaningful numbers or rate to characteristics of objects, persons, states or events according to pre-specified rules.

**Median age** The age that divides a population into two numerically equal groups; that is, half the people are younger than this age and half are older.

**Median family income** The level of income that divides family households into two numerically equal groups, with half the households above this income level and half below.

**Median household income** The level of income that divides households into two numerically equal groups, with half the households above this income level and half below.

**Medicaid** The federally sponsored and state-administered government health insurance program for the low-income population in the US.

**Medical care** The provision of formal clinical services to individual patients; medical care reflects a narrow notion of health and healthcare in contrast to health care.

**Medicare** The federally-sponsored health insurance program that provides health insurance coverage for the elderly population in the US.

**Metropolitan Statistical Area (MSA)** An official federal designation for a large concentration of population that includes a central city, a central county and the surrounding interrelated counties.

**Micropolitan statistical area** An official federal designation for an urban cluster of at least 10,000 but less than 50,000 population plus adjacent counties having a high degree of social and economic integration.

**Migration** The process whereby individuals or groups move from one geographic area to another with the purpose of permanent residence.

**Migration, internal** The movement of individuals in a population within a country's boundaries for purposes of changing residency; in the US a mover must cross a county line to be designated as a migrant.

**Migration, international** The movement of individuals within a population from one country to another country for purposes of permanent relocation.

**Migration, interstate** In the US, the movement of internal migrants across a state line for purposes of changing residency.

**Migration, intrastate** In the US, the movement of internal migrants within a particular state for purposes of changing residency; must include movement across a county line.

**Migration, net** The metric for measuring the level of migration that subtracts the number of out-migrants from the number of in-grants to generate a figure for net migration.

**Migration efficiency** A measure of migration calculated by dividing an area's net migration (in-migrants minus out-migrants) by gross migration (sum of in-migrants and out-migrants).

**Migration rate** One of several indicators developed to measure the level of population movement (either internally or internationally) into and out of a particular geographic area.

**Migration stream** A process where a large number of migrants or immigrants move as a group from one geographic area to another.

**Mobility, geographic** The extent to which members of a population move from place to place within a specific time period.

**Mobility, social** A change in social status typically referring to movement up or down a society's social hierarchy.

**Model** A statement (mathematical or otherwise) that specifies the conceptual relationship between two or more variables.

**Morbidity** The level of sickness and disability within a population, usually expressed in terms of incidence and prevalence rates.

**Mortality** The level of death characterizing a population, usually expressed in terms of the number of deaths and/or death rates.

**Mortality, age-adjusted** A mortality or death rate that has been statistically adjusted to standardize the population in order to make comparisons between two disparate populations possible.

**Mortality, age-specific** The mortality or death rate associated with various age cohorts.

**Mortality, sex-specific** The mortality or death rate associated with males and females, respectively.

**National Center for Health Statistics (NCHS)** The "census bureau" of health data, the NCHS is the federal agency responsible for the collection, management and dissemination of most national data on health and healthcare.

**Natural region** A geographical area that is characterized by homogenous social, economic and political attributes (e.g., the Mississippi "delta").

**Notifiable disease** A disease whose presence must be reported to public health officials; typically, an infectious disease.

**Outcome** In research, the observed effect that an intervention has on the characteristics of a population.

**Outcome measurement** A formal process for assessing the effectiveness of an intervention on the condition of a study population.

**Out-migration** The flow of residents out of a geographic area; technically, movers must move across county lines in order to be considered an out-migrant.

**Parity** The number of children previously born alive to a woman at the time of her current pregnancy.

**Policy** Authoritative decisions made in the legislative, executive or judicial branches of government (public policies) intended to direct or influence the actions, behaviors or decisions of others. Policies are also set within organizations (e.g., policies and procedures).

**Political or administrative unit** A bounded geographic area formally defined for administrative purposes, such as a state, county, municipality, or school district.

**Population at risk** The total number of persons within a population who are at risk for a particular condition at a specified point in time.

**Population decrease** The total amount of decline in a population resulting from the interaction of births, deaths, and migration for a given period of time.

**Population density** The resident population per unit of land area; the persons per square mile within a particular geographic area.

**Population distribution** The patterns of settlement and dispersal of a population within a defined geographic area.

**Population increase** The total amount of growth in a population resulting from the interaction of births, deaths, and migration in a given period of time. "Population decrease" is the antithesis of population increase.

**Population momentum** The tendency for population growth to continue beyond the time that replacement-level fertility has been achieved because of the relatively high concentration of people in the childbearing years.

**Population policy** Explicit or implicit measures instituted by a government to influence population size, growth, distribution, or composition.

**Population pyramid** A bar chart, arranged vertically, that shows the distribution of a population by age and sex. By convention, the younger ages are at the bottom, with males on the left and females on the right.

**Population registry** A government data collection system in which the location and demographic characteristics of a population are continuously recorded.

**Predictive modeling** A technique for predicting the number of people within a specified population who will have specified characteristics are some point in the future.

**Poverty** The financial state of a population that places it under economic stress; measured in the US in absolute terms (e.g., dollar amount) or in relative terms (i.e., relation to others in the population).

**Prediction** A figure generated in order to estimate the size and/or characteristics of a defined population at a future point in time based on the extrapolation of historical data.

**Prevalence** The total number of cases of a disease, disability, or other health-related condition at a particular point in time; used to calculate a prevalence rate.

**Primary data** Data generated directly through surveys, focus groups, observational methods, and other techniques.

**Primary research** A type of research that involves the direct collection of data for a specific use.

**Projection, population** A calculated figure indicating the size of the population for a particular area or population category for some point in the future.

**Psychographic (or lifestyle) segmentation** The process of subdividing a population into groups of like individuals on the basis of their psychographic attributes.

**Psychographics** Subjective information reported by people regarding their beliefs, feelings, attitudes and behavior patterns; also referred to as lifestyles and utilized as a basis for predicting behavior.

**Public health** The set of activities designated for public health agencies that include disease surveillance, health status monitoring, air and water monitoring, food inspection, the registration of vital events, and the control of infectious conditions.

**Qualitative research** Inductive, subjective and process-oriented research methods utilized to describe, understand, and interpret observed phenomena.

**Quantitative research** Deductive and objective research methods primarily designed to identify relationships between variables.

**Race** A term used to distinguish different populations based on their physical characteristics (most often skin color in the US); the racial categories identified by the US Census Bureau are: white, African-American (or Black); Asian and Pacific Islander, and American Indian.

**Rate** The level of occurrence of a phenomenon per a specified number of persons—e.g., per 100, 1000, 10,000 or 100,000.

**Ratio** The proportion of a characteristic in relation to another characteristic—e.g., a sex ratio of 95 means there are 95 males per 100 females.

**Reality-based estimation method** A methodology that generates estimates or projections for a defined population based on relevant proxy measures (such as housing units, utility hookups, or school enrollment).

**Reality-based population estimate** A figure generated on the size of a population based on proxy or symptomatic data as a basis for calculating an estimate or projection.

**Refugee** An individual who has left his country of origin and seeks admission to another country based on claims of danger, persecution or other adverse conditions in the country of origin; an officially stateless person recognized by the United Nations.

**Region, geographic** A geographic area that extends beyond any one political jurisdiction to create an internally consistent region of some type (e.g.,

Appalachia); one of four officially designated geographic areas into which the United States is divided by the federal government.

**Registry** The systematic recording and reporting of events or situations characterizing a population—e.g., a registry containing all reported cases of cancer within a specified area or all licensed physicians within a particular state.

**Replacement–level fertility** The level of fertility at which women produce enough births for the population to replace itself—generally 2.1 births per woman.

**Reproduction rate** See fertility rate.

**Respondent** A person or other unit (e.g., household) who responds to a survey instrument.

**Risk** The exposure that individuals or population groups experience with regard to various conditions.

**Rural area** An area designated by the Census Bureau for data collection purposes that does not meet the minimum standards for an urban area.

**Sample** A subset of the population selected for inclusion in a sample survey.

**Sample size** The number of individuals or other units selected for inclusion in a survey based on specified rules for determining the number.

**Sample survey** A survey in which a subset of the population has been selected for participation on the assumption that conclusions concerning the total population can be drawn based on a sample of respondents.

**Sampling frame** The list of the target population from which a sample will be drawn—e.g., all addresses in a particular community.

**Secondary data** Data collected for some other purpose (such as routine administrative records) but used for some research application–e.g., census data collected for federal statistical purposes used for market research.

**Segment** A component of a population or market defined on the basis of some characteristic relevant to analysts.

**Segmentation** The process of dividing a population into meaningful segments for purposes of analysis.

**Segmentation, demographic** The grouping of individuals into population segments based on such demographic characteristics as age, sex, income or race.

**Segmentation, geographic** The grouping of individuals into population segments based on location of residence or work.

**Segmentation, psychographic** The grouping of individuals into population segments based on lifestyle and attitudinal characteristics.

**Service area** The geographic area from which an organization draws the majority of its customers; often used interchangeably with "market area".

**Sex ratio** An indicator of the relative proportions of males and females within a population; typically calculated in terms of the number of males per100 females.

**Sex-specific mortality** The death rate for males and females for overall mortality and cause-specific mortality.

**Social determinants of health** Characteristics of society that contribute to conditions of health or illness within a population, thereby mitigating the effects of disease pathology and genetics.

**Sociocultural characteristic** A demographic attribute that reflect one's position in the social structure such as marital status, income, education and occupation.

**Spatial analysis** An analytical technique that involves the study of the spatial relationships between various phenomena—e.g., the analysis of the geographic distribution of people living in poverty.

**Stable population** A population with an unchanging rate of growth and an unchanging age composition as a result of age-specific birth and death rates that have remained constant over a sufficient period of time.

**Standardization** A process through which dissimilar populations are statistically adjusted to allow for meaningful comparisons.

**Statistical unit** A bounded geographic area formally defined for data collection purposes, such as the geographic units developed by the Census Bureau.

**Suburb** A residential area within an urban area located on the outskirts of the central city.

**Survey** A data collection technique that involves the use of a questionnaire administered in any number of ways.

**Survey instrument** Questionnaires used in survey research are referred to as survey instruments.

**Survival rate** The proportion of persons in a specified group (age, sex, or health status) alive at the beginning of an interval (such as a five-year period) who survive to the end of the interval.

**Synthetic data** Population estimates and projections that are generated using statistical techniques and models in the absence of actual data.

**Total Fertility Rate (TFR)** The average number of children who would be born to a woman (or group of women) during her lifetime if she were to pass through her childbearing years conforming to the age-specific fertility rates of a base year.

**Trend analysis** The longitudinal analysis of data for the purpose of identifying the presence or absence of consistency over time; a form of time series analysis.

**Urban** In the US a community or settlement with a population of 2000 or more with a specified population density.

**Urbanization** The process through which individuals within a population become more concentrated in urban areas than in rural areas.

**Uninsured** Individuals or groups in a population who are not covered under any health insurance plan; in financial accounting also referred to as self-pay.

**Urbanized area** A Census Bureau designation for a geographic area that meets specific requirements for being designated as "urban".

**Variable** A concept of interest in a research study, operationalized in a manner that will allow for statistical analysis.

**Vendor, commercial data** A private sector corporation that generates and disseminates data and/or software for data manipulation.

**Vital statistics** Data collected by government agencies related to "vital events"—e.g., births, deaths, marriages, divorces, and abortions.

**Years of potential life lost** A measure of disability and premature death that is calculated in terms of the number of years that an individual would have expected to live (or to have quality of life) had it not been for disability or death.

**ZIP Code** An administrative district created by the US Postal Service for the efficient delivery of mail. Although not originally intended as a unit for data collection and dissemination, ZIP codes are often used as a basis for planning and marketing studies.

# Index





R. K. Thomas, *Demography: An Introduction to Population Studies*, Springer Texts in Social Sciences, https://doi.org/10.1007/978-3-031-56623-3

**D**

MIX
Papier aus verantwortungsvollen Quellen
Paper from responsible sources
FSC® C105338

If you have any concerns about our products,
you can contact us on
**ProductSafety@springernature.com**

In case Publisher is established outside the EU,
the EU authorized representative is:
**Springer Nature Customer Service Center GmbH**
**Europaplatz 3, 69115 Heidelberg, Germany**

Printed by Libri Plureos GmbH
in Hamburg, Germany